Advances in Liquid Crystal Optical Devices

Advances in Liquid Crystal Optical Devices

Editors

Zhenghong He
Yuriy Garbovskiy

Basel • Beijing • Wuhan • Barcelona • Belgrade • Novi Sad • Cluj • Manchester

Editors
Zhenghong He
Southwest University
Chongqing
China

Yuriy Garbovskiy
Central Connecticut State University
New Britain
USA

Editorial Office
MDPI
St. Alban-Anlage 66
4052 Basel, Switzerland

This is a reprint of articles from the Special Issue published online in the open access journal *Crystals* (ISSN 2073-4352) (available at: https://www.mdpi.com/journal/crystals/special_issues/liquid_crystal_optical_device).

For citation purposes, cite each article independently as indicated on the article page online and as indicated below:

Lastname, A.A.; Lastname, B.B. Article Title. *Journal Name* **Year**, *Volume Number*, Page Range.

ISBN 978-3-7258-0845-8 (Hbk)
ISBN 978-3-7258-0846-5 (PDF)
doi.org/10.3390/books978-3-7258-0846-5

Cover image courtesy of Yuriy Garbovskiy

© 2024 by the authors. Articles in this book are Open Access and distributed under the Creative Commons Attribution (CC BY) license. The book as a whole is distributed by MDPI under the terms and conditions of the Creative Commons Attribution-NonCommercial-NoDerivs (CC BY-NC-ND) license.

Contents

About the Editors vii

Preface ix

Miyuki Harada and Takuya Matsumoto
Thermal Conductivity and Orientation Structure of Liquid Crystalline Epoxy Thermosets Prepared by Latent Curing Catalyst
Reprinted from: *Crystals* **2024**, *14*, 47, doi:10.3390/cryst14010047 1

Tomoki Shigeyama, Kohsuke Matsumoto, Kyohei Hisano and Osamu Tsutsumi
Angular-Dependent Back-Reflection of Chiral-Nematic Liquid Crystal Microparticles as Multifunctional Optical Elements
Reprinted from: *Crystals* **2023**, *13*, 1660, doi:10.3390/cryst13121660 13

Oleksandr V. Kovalchuk, Tetiana M. Kovalchuk and Yuriy Garbovskiy
Eliminating Ambiguities in Electrical Measurements of Advanced Liquid Crystal Materials
Reprinted from: *Crystals* **2023**, *13*, 1093, doi:10.3390/cryst13071093 22

Xinyue Zhang and Kun Li
Phase-Only Liquid-Crystal-on-Silicon Spatial-Light-Modulator Uniformity Measurement with Improved Classical Polarimetric Method
Reprinted from: *Crystals* **2023**, *13*, 958, doi:10.3390/cryst13060958 37

Wenbin Feng, Zhiqiang Liu, Hao Liu and Mao Ye
Design of Tunable Liquid Crystal Lenses with a Parabolic Phase Profile
Reprinted from: *Crystals* **2023**, *13*, 8, doi:10.3390/cryst13010008 55

Idriss Moundoungou, Zohra Bouberka, Guy-Joël Fossi Tabieguia, Ana Barrera, Yazid Derouiche, Frédéric Dubois, et al.
End-of-Life Liquid Crystal Displays Recycling: Physico-Chemical Properties of Recovered Liquid Crystals
Reprinted from: *Crystals* **2022**, *12*, 1672, doi:10.3390/cryst12111672 64

Mattia Sabadin, Jeroen A. H. P. Sol and Michael G. Debije
Direct Ink Writing of Anisotropic Luminescent Materials
Reprinted from: *Crystals* **2022**, *12*, 1642, doi:10.3390/cryst12111642 81

Dong-Wook Lee, Da-Bin Yang, Dong-Hyun Kim, Jin-Young Oh, Yang Liu and Dae-Shik Seo
Anisotropic Surface Formation Based on Brush-Coated Nickel-Doped Yttrium Oxide Film for Enhanced Electro-Optical Characteristics in Liquid Crystal Systems
Reprinted from: *Crystals* **2022**, *12*, 1554, doi:10.3390/cryst12111554 89

Magdalena Urbańska and Mateusz Szala
Synthesis, Mesomorphic Properties and Application of (R,S)-1-Methylpentyl 4′-Hydroxybiphenyl-4-carboxylate Derivatives
Reprinted from: *Crystals* **2022**, *12*, 1710, doi:/10.3390/cryst12121710 101

Le Zhou and Sijie Liu
Development and Prospect of Viewing Angle Switchable Liquid Crystal Devices
Reprinted from: *Crystals* **2022**, *12*, 1347, doi:10.3390/cryst12101347 115

About the Editors

Zhenghong He

Zhenghong He earned his B.S. degree in applied physics from Jianghan University in Wuhan, China, followed by an M.S. degree in Theoretical Physics from Huazhong University of Science and Technology in Wuhan, China. He later obtained his PhD in Electrical Engineering from Shanghai Jiaotong University in Shanghai, China. Currently, he holds the position of Associate Professor at the School of Physical Science and Technology at Southwest University in Chongqing, China. His research focuses on liquid crystals, magnetic materials, nanotechnology, and physics education.

Yuriy Garbovskiy

Yuriy Garbovskiy received his PhD degree in physics from the Institute of Physics (Kyiv, Ukraine). He is an associate professor at the Department of Physics and Engineering Physics at Central Connecticut State University, USA. His research interests include soft matter physics with an emphasis on liquid crystals, materials science and nanotechnology, optics and advanced optical materials, applied physics, biophysics, and physics education. He has authored over 90 research publications, one book published by Cambridge University Press, and one US patent. He is a Senior Member of OPTICA (formerly the Optical Society of America). He has edited several special issues of academic journals and currently serves on the editorial boards of multiple peer-reviewed journals, such as *Applied Optics* (until 2022), *Optics* (since 2019), and *Frontiers in Soft Matter* (since 2022).

Preface

We are witnessing the constant evolution and rapid development of liquid crystal science and technology. Since making a revolution in the display industry, liquid crystals have become an important part of a human civilization. The tunability of liquid crystal material properties successfully exploited in liquid crystal displays has also enabled a plethora of their non-display applications, including numerous electro-optical components (filters, shutters, waveplates, lenses, and waveguides), reconfigurable microwave devices (antennas, phase shifters, and delay lines), sensors, transducers, and actuators including liquid crystal-based soft robotics, miniature lasers and intensity modulators, large privacy windows, and many other tunable optical and electro-optical devices.

Given the rapid growth and expansion of both display and non-display applications of liquid crystals, it is important for scientists to stay abreast of recent developments in the field. The present Special Issue has been created to realize this need. The ten papers in this Special Issue, which are authored by 37 authors from 10 different countries, will immediately give the reader an insight into the multidisciplinary and collaborative nature of liquid crystal research. The covered topics are as diverse as viewing angle controllable liquid crystal devices, the synthesis of new liquid crystal materials, the recovery of liquid crystal materials from recycled displays, the alignment of liquid crystals by unconventional surfaces, the use of liquid crystal elastomers for direct ink writing, electrical measurements of liquid crystals, phase-only liquid-crystal-on-silicon spatial-light-modulator uniformity measurements, the design of tunable liquid crystal lenses, chiral-nematic liquid crystal microparticles as multifunctional optical elements, and heat dissipation properties of liquid crystalline epoxies. These papers provide a snapshot of the latest research developments and offer some insights into future progress in the field. We believe that this Special Issue will be of high interest to any reader interested in liquid crystals and their applications.

This Special Issue would not be possible without the contribution of its authors, anonymous reviewers, Editorial Board Members, and the editorial staff members of *"Crystals"* to whom we are very grateful.

Zhenghong He and Yuriy Garbovskiy
Editors

Article

Thermal Conductivity and Orientation Structure of Liquid Crystalline Epoxy Thermosets Prepared by Latent Curing Catalyst

Miyuki Harada * and Takuya Matsumoto

Department of Chemistry and Materials Engineering, Faculty of Chemistry, Materials and Bioengineering, Kansai University, Osaka 564-8680, Japan
* Correspondence: mharada@kansai-u.ac.jp

Abstract: Improvements in the performance of electronic devices necessitate the development of polymer materials with heat dissipation properties. Liquid crystalline (LC) epoxies have attracted attention because of the orientation of their polymer network chains and their resultant high thermal conductivity. In this study, a diglycidyl ether of 1-methyl-3-(4-phenylcyclohex-1-enyl)benzene was successfully synthesized as an LC epoxy and the LC temperature range was evaluated via differential scanning calorimeter (DSC). The synthesized LC epoxy was cured with m-phenylenediamine (m-PDA) as an amine-type curing agent and 1-(2-cyanoethyl)-2-undecylimidazole (CEUI) as a latent curing catalyst, respectively. The LC phase structure and domain size of the resultant epoxy thermosets were analyzed through X-ray diffraction (XRD) and polarized optical microscopy (POM). High thermal conductivity was observed in the m-PDA system (0.31 W/(m·K)) compared to the CEUI system (0.27 W/(m·K)). On the other hand, in composites loaded with 55 vol% Al_2O_3 particles as a thermal conductive filler, the CEUI composites showed a higher thermal conductivity value of 2.47 W/(m·K) than the m-PDA composites (1.70 W/(m·K)). This difference was attributed to the LC orientation of the epoxy matrix, induced by the hydroxyl groups on the alumina surface and the latent curing reaction.

Keywords: epoxy; thermosets; crosslinking; mechanical properties; thermal property; liquid crystals

Citation: Harada, M.; Matsumoto, T. Thermal Conductivity and Orientation Structure of Liquid Crystalline Epoxy Thermosets Prepared by Latent Curing Catalyst. *Crystals* **2024**, *14*, 47. https://doi.org/10.3390/cryst14010047

Academic Editors: Zhenghong He and Yuriy Garbovskiy

Received: 28 November 2023
Revised: 25 December 2023
Accepted: 26 December 2023
Published: 28 December 2023

Copyright: © 2023 by the authors. Licensee MDPI, Basel, Switzerland. This article is an open access article distributed under the terms and conditions of the Creative Commons Attribution (CC BY) license (https://creativecommons.org/licenses/by/4.0/).

1. Introduction

In recent years, as electronic devices have become more sophisticated and densely packaged, the amount of heat dissipated inside them has tended to increase. This causes functional failure and deterioration of peripheral components. Therefore, a need for higher thermal conductivity exists for encapsulating materials [1]. Epoxy resins are one of the conventional thermosetting polymers. They have been widely used in industrial fields, such as adhesives, coatings, and composite materials for their excellent bonding and thermal properties, chemical resistance, low shrinkage, processability, and electrical insulation properties. Therefore, epoxy resins are also used as encapsulation materials and printed circuit boards; however, general-purpose epoxy resins are known to have lower thermal conductivity than metallic and ceramic materials. In general, inorganic filler powders, such as alumina (Al_2O_3) [2–4] and boron nitride (BN) [5–12], are added to the resin matrix to attain encapsulating materials with high thermal conductivity. However, the extent to which composite fillers can improve thermal conductivity is limited because the increase in filler content causes matrix embrittlement and deterioration of molding processability because of high viscosity. Therefore, an urgent need exists to increase the thermal conductivity of the epoxy resin itself, which has been reported to further increase the thermal conductivity when it is filled with a filler.

Liquid crystalline (LC) epoxies with mesogen groups in the backbone moiety have received attention. Frequently, they contain azo, azomethine, ester, stilbene, biphenyl, and terphenyl as mesogenic groups. LC epoxies form LC structures such as nematic (N) and smectic (S_m) phases via π–π stacking of mesogen groups. Not only in the monomeric state but also upon reaction with a curing agent, LC epoxies form LC domains in networked polymer chains. This oriented polymer structure efficiently suppresses phonon scattering and demonstrates superior thermal conductivity compared to general-purpose epoxy thermosets [13–34]. Akatsuka et al. have reported that an ester-type twin mesogen epoxy thermoset exhibits a thermal conductivity approximately four to five times higher than those of general-purpose epoxy resins [8]. Tokushige et al. reported that cured materials that form an S_m phase exhibit higher thermal conductivity than those that form an N phase [28]. However, in general, LC epoxy resin monomers with mesogen groups have a high melting point; thus, high-temperature conditions are critical during the preparation of cured products. However, high-temperature conditions have disadvantages, including high energy costs in industrial mass production. To solve this problem, we synthesized a phenyl-cyclohexene type LC epoxy with a low melting point and wide LC temperature range and subsequently found that the resultant thermosets exhibited high thermal conductivity [15].

Another possible curing method to improve LC alignment is the self-polymerization reaction of LC epoxy resins using a latent curing catalyst. Tertiary amine compounds and imidazole compounds are used as curing catalysts in the self-polymerization reaction of epoxy resins. These compounds are added to the epoxy resin in small amounts (1–5 phr). The cured LC epoxy thermosets prepared with a curing catalyst have a high concentration of mesogenic groups in their networked structure. However, there are imidazoles with a cyano group (–CN) [35] as a strong electron-withdrawing group. In addition, because of the lower electron density, imidazoles with a cyano group exhibit lower reactivity than conventional imidazole compounds. LC epoxy resins prepared with less-reactive curing agents have longer gelation times and can provide a range of possible mesogenic groups. We have also reported on the improvement of the alignment and toughness of cured materials using a mixture of LC epoxy resin and a rigid, highly reactive curing agent, where the curing agent was present in small amounts, contained flexible chains, and exhibited low reactivity [36]. The use of latent curing catalysts increased the concentration of mesogen groups in the LC epoxy thermosets and extended the gelation time, giving the mesogen groups more time to align; this alignment is expected to improve the orientation of networked polymer chains.

In the present study, a diglycidyl ether of 1-methyl-3-(4-phenylcyclohex-1-enyl)benzene as an LC epoxy monomer was synthesized and its LC property and reactivity were investigated. The differences in the LC phase structure and the thermal conductivity of the cured materials prepared using aromatic diamines as a curing agent and imidazoles as a latent curing catalyst were discussed. These properties were also compared with those of epoxy thermosets prepared with a conventional aromatic amine. The thermal conductivity of composites loaded with alumina particles as a high thermally conductive filler was also investigated.

2. Materials and Methods

2.1. Materials

The epoxy resin, curing agent, and curing catalyst used were diglycidyl ether of 1-methyl-3-(4-phenylcyclohex-1-enyl)benzene (DGEDPC-Me) (M_w = 392 g/mol, Cr 80 LC 132 Iso), m-phenylenediamine (m-PDA, M_w = 108 g/mol, m.p. 62–65 °C, Wako Pure Chemical Industries, Osaka, Japan), and 1-(2-cyanoethyl)-2-undecylimidazole (CEUI, M_w = 275 g/mol, m.p. 49–54 °C, Tokyo Kasei Chemical Industry, Tokyo, Japan), respectively. The structures of these chemicals are shown in Figure 1.

Al_2O_3 particles (AA-05, Sumitomo Chemical, particle mean diameter: 5 μm) were used as a thermal conductive filler.

Hexamethyl disilazane (HMDS, M_w = 275 g/mol, Tokyo Kasei Chemical Industry, Tokyo, Japan) was used as a silane decoupling agent for a glass substrate.

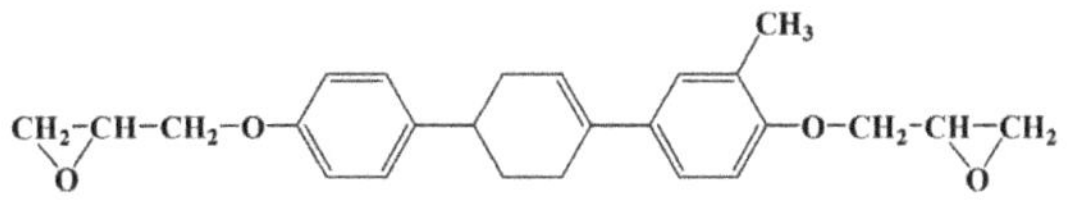

Diglycidyl ether of 1-methyl-3-(4-phenylcyclohex-1-enyl)benzene (DGEDPC-Me)

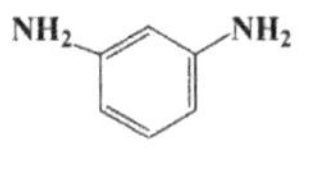

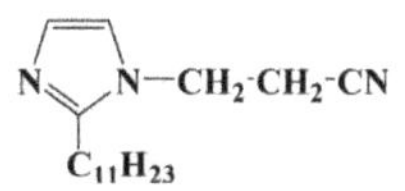

m-Phenylenediamine (m-PDA)   1-(2-Cyanoethyl)-2-undecylimidazole (CEUI)

Figure 1. Chemical structures of DGEDPC-Me, m-PDA, and CEUI.

2.2. Synthesis of 1,4-Diphenyl-3-Methyl-Cyclohexene-Type Epoxy Resin (DGEDPC-Me)

The DGEDPC-Me synthesis pathway is based on that reported in our previous study [32] on the same structure without the methyl branch. DGEDPC-Me was synthesized from a 24-times excess of 1-chloro-2,3-epoxypropane (119.64 g, 12.9 × 10^2 mmol, M_w = 93 g/mol; Wako Pure Chemical Industries, Osaka, Japan) and 4,4′-dihydroxy-(1-methyl-3-(4-phenylcyclohex-1-enyl)benzene) (15.00 g, 53.6 mmol, M_w = 280 g/mol; Honshu Chemical Industry, Tokyo, Japan) with 60 mL of dimethyl sulfoxide (DMSO) and tetra-n-butylammonium chloride (TBAC) (90.0 mg, 32.4 × 10^{-3} mmol, M_w = 278 g/mol, Sigma-Aldrich Co. LLC, St. Louis, MO, USA) as the catalyst. The epoxidation reaction was conducted at 60 °C for 1 h, after which a 50 wt% aqueous sodium hydroxide (M_w = 40 g/mol, Wako Pure Chemical Industries, Osaka, Japan) solution (5.15 g, 129 mmol) was added over 0.5 h and additionally reacted for 1.5 h to induce the epoxy ring-closure reaction. The product was purified through reprecipitation with 750 mL of methanol, followed by 5 washes with 20 mL of methanol. It was then dried at 60 °C for 2 h. The obtained product contained 15.69 g of white powder, and the reaction yield was 75%. All reagents in the above synthesis were used as received, without any purification. The structure of the synthesized compounds was confirmed using ^{1}H-NMR (JNM-AL 400; JEOL Ltd., Tokyo, Japan) and Fourier transform infrared (FT-IR, SPECTRUM 100, Perkin Elmer, Waltham, MA, USA) spectroscopy. The repeating unit was determined to be 0.06 via gel permeation chromatography (GPC, LC20AD, Shimadzu Corp., Kyoto, Japan) with THF as the eluent at 40 °C, using a refractive index detector (RID-20A, Shimadzu Corp., Kyoto, Japan). The sample (about 5 mg) was completely dissolved in 10 mL of THF.

^{1}H-NMR and FT-IR spectra are shown in Supplementary Materials Figures S1 and S2.

^{1}H-NMR (CDCl$_3$, δ, ppm): 1.9 (q, 2H, CH$_2$), 2.1 (d, 2H, CH$_2$), 2.3 (s, 3H, CH$_3$), 2.5 (t, 2H, CH$_2$), 2.8 (m, 1H, CH), 2.9 (d, 4H, CH$_2$, epoxy), 3.4 (m, 2H, CH, epoxy), 3.9 (d, 2H, CH$_2$, epoxy), 4.1 (d, 2H, CH$_2$, epoxy), 6.1 (s, 1H, CH), 6.7 (d, 2H, CH, aromatic), 6.9 (d, 2H, CH, aromatic), 7.1 (d, 2H, CH, aromatic), 7.2 (d, 2H, CH, aromatic), 7.3 (d, 2H, CH, aromatic).

IR (KBr, cm^{-1}): 3630–3095 (ν O–H), 3090–2900 (ν C–H, methylene), 2900–2790 (ν C–H, methyl), 1605, 1510 (ν C=C, aromatic), 1460 (δ C–H, methyl), 1390 (ν C–O, phenol), 1350 (δ C–H, methyl), 1030 (ν C–O, ether), 915 (ν C–O, epoxy).

The transition temperature of the synthesized DGEDPC-Me was measured with a differential scanning calorimeter (DSC Exter 7020, SII, Chiba, Japan) at a heating rate of 5 °C/min. The synthesized epoxy monomer was found to exhibit an LC phase between 80 and 132 °C (C 80 LC 132 I).

2.3. Curing of Epoxy Resin and Al$_2$O$_3$ Composites

DGEDPC-Me was cured with a stoichiometric amount (NH: epoxy group = 1:1) of the curing agent (m-PDA) or 3 phr of latent curing catalyst (CEUI). The epoxy resin was completely melted at 160 °C for 2 min, and *m*-PDA or CEUI was then added. Both systems were cured in an oven at 70 °C for 3 h, at 80 °C for 1 h, and at 100 °C for 1 h. They were then cured at 220 °C for 0.5 h. The heating rate was 20 °C/min. The composites, loaded with Al_2O_3 particles (particle mean diameter: 5 μm) without any surface treatment, were prepared using the same procedure as above.

2.4. Surface Treatment of a Glass Substrate

The surface of a glass substrate was directly treated with HMDS at room temperature through the dry treatment method.

2.5. Characterization Techniques

The structure of the synthesized epoxy monomers and the conversion of the epoxy groups in their thermosets were analyzed through FT-IR spectroscopy (SPECTRUM 100, Perkin Elmer, MA, USA), using samples incorporated into KBr pellets. The resolution of the FT-IR spectrum was 4 cm^{-1}, and the spectra were collected after four scans. The chemical conversion of the epoxy groups was determined from the reduced size of the peak at 910 cm^{-1}, which was identified as an epoxy group. Here, the peak at 1510 cm^{-1}, which was identified as a benzene ring, was used as an internal standard.

The gel fraction was estimated on the basis of the change in weight of the powdered bulk samples (0.50 g). The samples in cylindrical filter paper were soaked and stacked in THF at 25 °C for 1 h, then at 40 °C for 1 h, and then at 60 °C for 1 h. Following extraction, the samples were fully dried at 60 °C for 6 h under vacuum, and their weight was subsequently measured.

The phase transition behavior and LC texture of the synthesized epoxy monomers and their thermosets were confirmed under cross-polarized light using a polarized optical microscope (POM; BH-2, Olympus Corporation, Tokyo, Japan) equipped with a hot stage (TPC-2000, Ulvac, Inc., Kanagawa, Japan); the samples were heated at 5 °C/min. The thickness of the thermoset samples was 10 μm.

The LC phase orientation in the epoxy thermosets was confirmed using wide-angle X-ray diffraction (WAXD; NANO-Viewer MicroMax-007HF, Rigaku Corporation, Tokyo, Japan) and an imaging plate detector (R-AXIS-IV, Rigaku Corporation, Tokyo, Japan). The layer spacing, d, was calculated using Bragg's formula ($n\lambda = 2d \cdot \sin\theta$). Diffraction patterns were obtained using Cu Kα (λ = 0.145 nm) radiation generated at 40 kV and 30 mA. The thickness of the epoxy thermoset samples was 1.0 mm.

The thermal conductivity at 25 °C was determined from the thermal diffusivity α, density ρ, and specific heat capacity C_p. The thermal conductivity λ was calculated using the following equation [37]:

$$\lambda = \alpha \cdot \rho \cdot C_p \quad (1)$$

The thermal diffusivity α of the systems at 25 °C was measured using Xe-flash analysis (LFA 447 NanoFlash, Netzsch, Selb, Germany) according to the ISO 18755 standard. The samples were prepared as disks with a diameter of 10.0 mm and thickness of 1.0 mm. The entire surface of the samples was coated with a Au layer of 500 Å via ion sputtering and then covered with a carbon layer of ~8 μm using a carbon sprayer (graphite coat; Nihon Senpaku) to enhance the thermal contact and prevent the direct transmission of light from the laser light source throughout the specimen. The measurement temperature was 25 °C.

The density ρ of the systems was measured at 25 °C using the pycnometer method according to the standard JIS K7112. The C_p was measured using a differential scanning calorimeter (DSC Exster 7020, SII, Chiba, Japan) at a heating rate of 5 °C/min. The sample weight was 10 mg, and the reference sample was α-Al_2O_3 (sapphire) (C_p: 0.78 J/(g·K)).

3. Results

3.1. Curing Reaction of the DGEDPC-Me/m-PDA and CEUI Systems

The conversion of the epoxy groups in the DGEDPC-Me systems was calculated through FT-IR measurements to confirm the effect of aromatic amine curing agents or latent catalysts on reactivity (Figure 2). The results showed that the reactivity of the epoxy group in the aromatic amine *m*-PDA system was consumed rapidly, whereas the latent catalyst CEUI system showed a substantially lower reactivity in the initial stage of curing, and its reactivity gradually increased with increasing curing temperature. We speculate that the electron-withdrawing behavior of the cyano group of CEUI strongly suppressed the reactivity of the epoxy group.

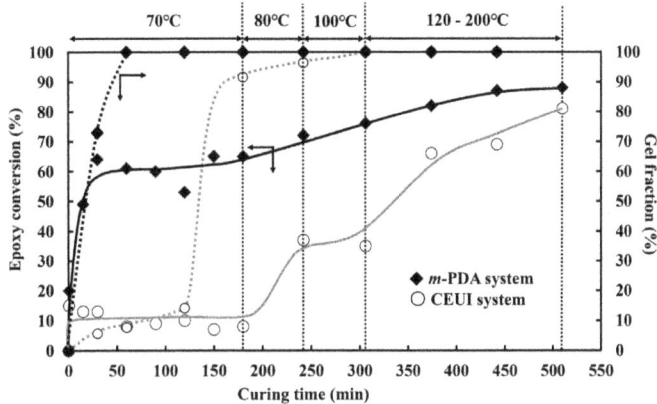

Figure 2. Epoxy conversion (solid line) and gel fraction (dotted line) of the (♦) *m*-PDA and (○) CEUI systems.

To confirm that the latent catalyst affects the extension of the gelation time, the gel fraction was measured during the curing process. As a result, in the CEUI system, extension of the gelation was observed with the latent catalyst, suggesting that the mesogenic groups had sufficient time for arrangement.

3.2. LC Phase Structure of the DGEDPC-Me/m-PDA and CEUI Systems

Figure 3 shows polarized optical micrographs collected during the curing process to investigate the differences in the liquid crystallinity of the DGEDPC-Me systems. The aromatic amine *m*-PDA system showed a dark field throughout the region at a curing time of 30 s immediately after melting, and an isotropic phase was formed temporarily. As the curing time progressed, a liquid-crystalline phase appeared after ~3 min and the growth of the LC domains was observed until ~10 min. However, the latent catalyst CEUI system showed an LC phase after melting, without isotropization. In addition, the LC phase was maintained even as the curing time progressed, and the LC domain diameter increased. The LC domains of the *m*-PDA system were ~60 μm, whereas those of the CEUI system were ~120 μm, indicating that larger domains were formed in the latent catalyst system. This result is attributable to the higher concentration of LC epoxy resin in the system. The concentration of mesogenic groups in the CEUI system with 3 phr catalyst added is higher than that in the *m*-PDA system because the aromatic amine was added on a chemical-equivalent basis. This chemical-equivalent addition of the aromatic amine is speculated to have improved the stacking effect between the mesogen groups and prolonged the gelation time by suppressing the reaction rate, thereby allowing the mesogen groups to self-align more easily.

Figure 3. Polarized optical micrographs of the DGEDPC-Me thin films cured with (**a**) *m*-PDA and (**b**) CEUI systems during the curing process (curing temperature: 80 °C, magnification: ×200).

In addition, the bulk thermosets prepared in Al cups were polished to thin films and observed using POM (Figure 4). The birefringence pattern was confirmed in both systems under crossed Nicols, indicating that the LC phase was present in the DGEDPC-Me epoxy thermosets. When the crossed Nicols angle was rotated by 45°, the LC domains were identified as the area where bright and dark fields were switched. The results indicate that larger LC domains were formed in the CEUI (latent curing catalyst) system compared to those in the aromatic amine. A similar trend was confirmed in the bulk thermosets (Figure 3).

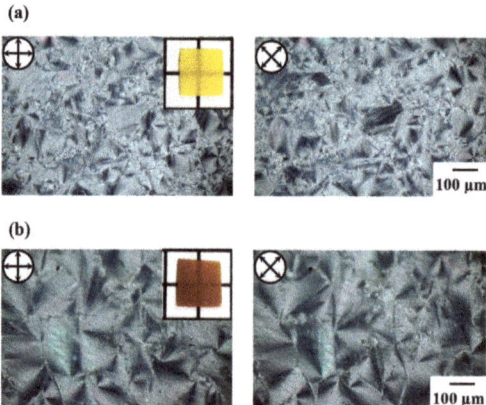

Figure 4. Polarized optical micrographs of the DGEDPC-Me cured with (**a**) *m*-PDA and (**b**) CEUI systems (magnification: ×100).

Figure 5 shows the XRD patterns acquired to identify the LC phase structure of the obtained LC epoxy thermosets cured through the m-PDA and CEUI. The results show a sharp peak at 2θ = 3.9° (2.26 nm) for (a) the *m*-PDA system and 4.4° (2.01 nm) for (b) the CEUI system on the small-angle side. Halos at 2θ = 18.4° (0.48 nm) were observed in the patterns of both systems. These results indicate that both systems formed an LC phase containing a S_m phase structure. The layer distance of the S_m phase formed by addition polymerization (*m*-PDA) and self-polymerization (CEUI) differed slightly. This result is attributable to the introduction of the amine curing agent by covalent bonding in the *m*-PDA system, resulting in a longer smectic layer distance than in the CEUI system. It reflects the difference in the assumed crosslinking structure. In addition, the relative intensity of the S_m (small-angle side) peak to the halo intensity, based on the nematic phase or amorphous structure, was calculated as shown in Table 1. The higher this value, the greater the extent of alignment of the mesogenic groups. The higher relative intensity for the *m*-PDA system indicates that more of the S_m phase was formed, whereas the CEUI system formed more of the nematic phase. In our previous paper, we reported on the LC epoxy thermosets, the halo strength of which is significantly lower than that of its smectic peak [36]. Therefore, we think that the *m*-PDA and CEUI systems consist of a mixture of smectic and nematic phases.

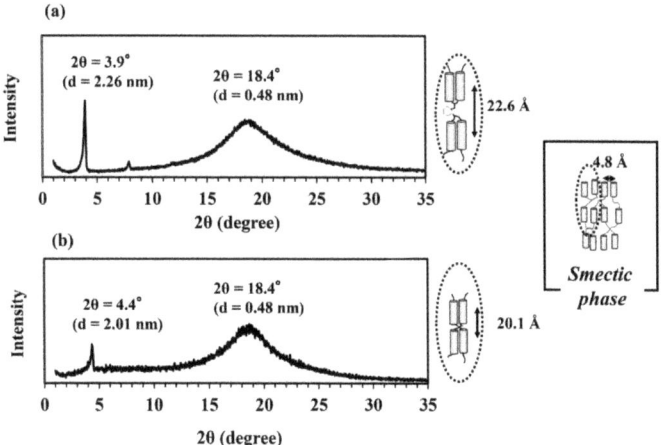

Figure 5. XRD patterns of the DGEDPC-Me cured with (**a**) *m*-PDA and (**b**) CEUI systems.

Table 1. Thermal conductivity of the DGEDPC-Me cured with the *m*-PDA and CEUI systems.

Curing System	Thermal Conductivity (W/(m·K))	Domain Size (μm)	Relative Intensity *
m-PDA system	0.31 ± 0.03	80	1.3
CEUI system	0.27 ± 0.02	120	0.7

* Calculated by $I_{3.9, 4.4°} / I_{Halo}$.

3.3. Thermal Conductivity of the DGEDPC-Me/m-PDA and CEUI Systems

The thermal conductivity of the LC epoxy thermosets cured through the m-PDA and CEUI is shown in Table 1. The results show that the *m*-PDA system exhibited a higher thermal conductivity (0.31 W/(m·K)) compared to the CEUI system (0.27 W/(m·K)). This higher thermal conductivity is attributed to the orientation structure of the networked polymer formed by the different curing reaction mechanisms. We have reported that the formation of the LC phase can be prevented with the loading of a BN filler [17]. However, Tanaka et al. reported that hydroxyl groups on the surface of glass substrates induced the formation of vertically oriented smectic phase structures when LC epoxy resin was cured with aromatic amine curing agents [25].

Here, the difference in the amount of hydroxyl groups in the formed networked polymer chains was investigated on the basis of the FT-IR measurements of both curing systems. More OH groups were formed in the *m*-PDA system than in the CEUI system. In the catalyst curing process, epoxy groups are consumed by the self-polymerization reaction derived from oxygen anions, resulting in an ether-linked network polymer. In this reaction, hydroxyl groups are hardly produced. However, the amine curing reaction generates one hydroxyl group per epoxy group because crosslinking occurs via addition polymerization of the amine and epoxy groups (Scheme 1). This difference in reaction mechanism results in a large difference in the amount of OH groups in these epoxy thermosets.

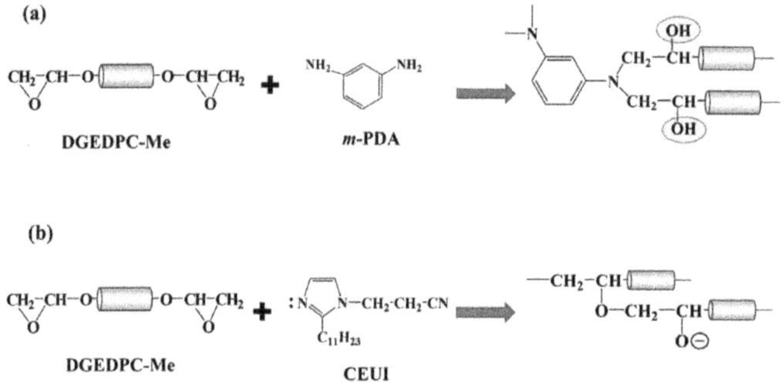

Scheme 1. Curing reaction of the DGEDPC-Me cured with (**a**) *m*-PDA and (**b**) CEUI systems.

3.4. Alignment of the DGEDPC-Me/m-PDA and CEUI Systems on Glass Substrates

We conducted a model experiment to investigate the influence of the amount of hydroxyl groups formed by the reaction on the orientation of mesogen groups in the thermosets. The *m*-PDA system, which forms hydroxyl groups via a crosslinking reaction, and the CEUI system, which generates almost no hydroxyl groups, were cured on a glass substrate with hydroxyl groups on its surface. We investigated whether the OH groups present on the outside induced the alignment of mesogen groups in the reticular chains of the CEUI system. Figure 6 shows the results of X-ray diffraction measurements of LC epoxy cured on a glass substrate. As a result, a sharp, high-intensity peak at $2\theta \approx 4°$ and a low-intensity halo at $2\theta \approx 18°$ were observed in the patterns of both the *m*-PDA and CEUI systems, indicating that, compared to the bulk thermosets, the CEUI system formed a larger amount of smectic phase structures. This result suggests that the presence of external hydroxyl groups leads to the alignment of mesogen groups in the CEUI system, which generates almost no hydroxyl groups.

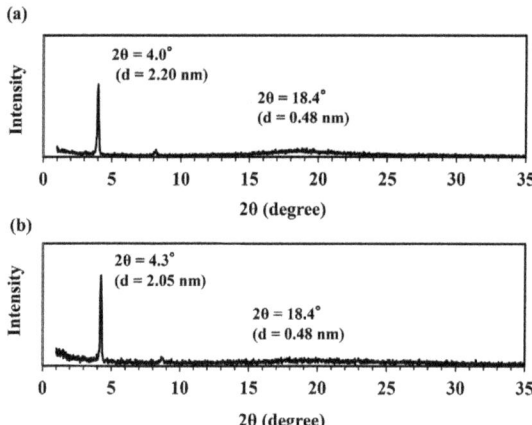

Figure 6. XRD patterns of the DGEDPC-Me cured with (**a**) *m*-PDA and (**b**) CEUI systems on glass substrates.

For further comparative study, we subjected glass substrates to a hydrophobic surface treatment using hexamethylene disilazane (HMDS). The DGEDPC-Me/CEUI system was cured on the hydrophobic glass substrates. The results of XRD measurements of the system are shown in Figure 7. Small-angle peaks derived from the smectic phase structure were identified. The system of the board showed lower values. This finding is attributable to the hydrophobic surface treatment substantially reducing the hydroxyl groups, making an arrangement of mesogenic groups difficult. These results confirmed that more hydroxyl groups on the substrate surface led to better mesogen group orientation. Tanaka et al. have reported that the LC epoxy orientation is induced by the formation of hydrogen bonds between the hydroxyl groups formed by the ring-opening reaction of epoxy groups and the hydroxyl groups on the surface of the glass substrate [25]. However, the results in Figures 6 and 7 indicate that the LC orientation is induced by interaction with the substrate even in the catalyst curing system, which generates a small amount of hydroxyl groups in the curing reaction.

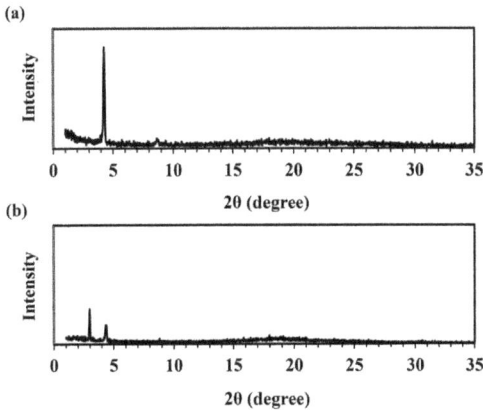

Figure 7. XRD patterns of CEUI systems cured on glass substrates. (**a**) Untreated and (**b**) treated with HMDS.

3.5. Thermal Conductivity of the DGEDPC-Me/m-PDA and CEUI/Al$_2$O$_3$ Composites

The DGEDPC-Me composites loaded with alumina particles having hydroxyl groups on the surface were prepared for the m-PDA and CEUI systems. On the basis of the results of the model experiment on the hydrophilic glass surface in Figure 7, we expected the S$_m$ phase orientation-induced effect of the mesogen groups in the networked chains to influence the surface of the alumina filler, especially in the CEUI system. Figure 8 shows the relationship between the thermal conductivity of the resultant composites and the volume fraction of Al$_2$O$_3$ particle. The results show that the thermal conductivity of both the m-PDA and CEUI systems improved dramatically with increasing addition of alumina particles. It is noteworthy that the CEUI composites loaded with 55 vol% Al$_2$O$_3$ particles showed a higher thermal conductivity value of 2.47 W/(m·K) than the m-PDA composites (1.70 W/(m·K)). A substantial improvement in thermal conductivity was observed in the CEUI system in particular, despite the lower thermal conductivity of the unloaded system. This result is attributable to the hydroxyl groups on the surface of the dispersed alumina particles improving the orientation of the mesogen groups in the CEUI system. In addition, the latent curing catalyst might have suppressed the initial curing reactivity, resulting in delayed gelation, providing sufficient time for orientation induction.

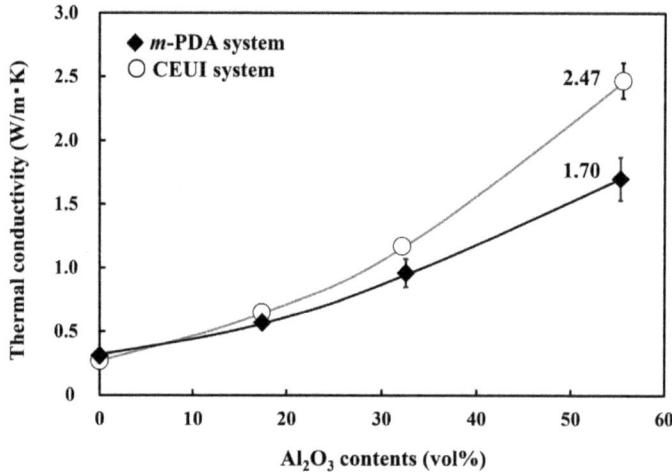

Figure 8. Thermal conductivity of the DGEDPC-Me cured with (♦) m-PDA and (○) CEUI systems loaded with Al_2O_3 (5 μm).

4. Conclusions

A diglycidyl ether of 1-methyl-3-(4-phenylcyclohex-1-enyl)benzene was cured using both m-PDA as an aromatic amine and CEUI as a latent curing catalyst. The CEUI catalyst system showed a long gelation time in the curing process and the obtained epoxy thermoset formed larger LC domains than the m-PDA system. However, the smectic LC orientation in the CEUI system was poorer than that in the m-PDA system. The thermal conductivity of the CEUI system was 0.27 W/(m·K), and the value was low compared to that of the m-PDA system (0.31 W/(m·K)). In addition, composites loaded with alumina particles were prepared for both the m-PDA and CEUI systems. Though the thermal conductivity of both systems improved with increasing addition of alumina particles, a substantial improvement in thermal conductivity was observed in the CEUI system. Even in the CEUI system, which had almost no hydroxyl groups, the delayed gelation and the hydroxyl groups on the alumina surface improved the mesogen group alignment and the thermal conductivity of the composite.

Supplementary Materials: The following supporting information can be downloaded at: https://www.mdpi.com/article/10.3390/cryst14010047/s1. Figure S1. ^{1}H-NMR spectra of the DGEDPC-Me (a) before and (b) after epoxidation; Figure S2. FT-IR spectra of the DGEDPC-Me (a) before and (b) after epoxidation.

Author Contributions: Conceptualization, methodology, formal analysis, investigation, writing original draft preparation, M.H.; data curation, M.H. and T.M.; writing—review and editing, M.H.; project administration, M.H. All authors have read and agreed to the published version of the manuscript.

Funding: This research received no external funding.

Data Availability Statement: The data presented in this study are available in this article and supplementary material.

Conflicts of Interest: The authors declare no conflicts of interest.

References

1. Lee, H.; Smet, V.; Tummala, R. Review of packaging schemes for power module. *IEEE J. Emerg. Sel. Top. Power Electron.* **2020**, *8*, 1.
2. Yetgin, H.; Veziroglu, S.; Aktas, O.; Yalçinkaya, T. Enhancing thermal conductivity of epoxy with a binary filler system of h-BN platelets and Al_2O_3 nanoparticles. *Int. J. Adhes. Adhes.* **2020**, *98*, 102540. [CrossRef]
3. Li, H.; Zheng, W. Enhanced thermal conductivity of epoxy/alumina composite through multiscale-disperse packing. *J. Compos. Mater.* **2021**, *55*, 17–25. [CrossRef]

4. Tian, F.; Cao, J.; Ma, W. Enhanced thermal conductivity and rheological performance of epoxy and liquid crystal epoxy composites with filled Al_2O_3 compound. *Polym. Test.* **2023**, *120*, 107940. [CrossRef]
5. Liu, Z.; Li, J.; Liu, X. Novel functionalized BN nanosheets/epoxy composites with advanced thermal conductivity and mechanical properties. *ACS Appl. Mater. Interfaces* **2020**, *12*, 6503–6515. [CrossRef] [PubMed]
6. Wang, Z.; Zhang, T.; Wang, J.; Yang, G.; Li, M.; Wu, W. The investigation of the effect of filler sizes in 3D-BN skeletons on thermal conductivity of epoxy-based composites. *Nanomaterials* **2022**, *12*, 446. [CrossRef] [PubMed]
7. Du, X.; Yang, W.; Zhu, J.; Fu, L.; Li, D.; Zhou, L. Aligning diamond particles inside BN honeycomb for significantly improving thermal conductivity of epoxy composite. *Compos. Sci. Technol.* **2022**, *222*, 109370. [CrossRef]
8. Yu, C.; Zhang, J.; Li, Z.; Tian, W.; Wang, L.; Luo, J.; Li, Q.; Fan, X.; Yao, Y. Enhanced through-plane thermal conductivity of boron nitride/epoxy composites. *Compos. Part A* **2017**, *98*, 25–31. [CrossRef]
9. Bian, W.; Yao, T.; Chen, M.; Zhang, C.; Shao, T.; Yang, Y. The synergistic effects of the micro-BN and nano-Al_2O_3 in micro-nano composites on enhancing the thermal conductivity for insulating epoxy resin. *Compos. Sci. Technol.* **2018**, *168*, 420–428. [CrossRef]
10. Yang, X.; Zhu, J.; Yang, D.; Zhang, J.; Guo, Y.; Zhong, X.; Kong, J.; Gua, J. High-efficiency improvement of thermal conductivities for epoxy composites from synthesized liquid crystal epoxy followed by doping BN fillers. *Compos. Part B Eng.* **2020**, *185*, 107784. [CrossRef]
11. Wu, L.; Huang, Y.; Yeh, Y.; Li, C. Characteristic and Synthesis of High-Temperature Resistant Liquid Crystal Epoxy Resin Containing Boron Nitride Composite. *Polymers* **2022**, *14*, 1252. [CrossRef] [PubMed]
12. Bao, Q.; He, R.; Liu, Y.; Wang, Q. Multifunctional boron nitride nanosheets cured epoxy resins with highly thermal conductivity and enhanced flame retardancy for thermal management applications. *Compos. Part A* **2023**, *164*, 107309. [CrossRef]
13. Akatsuka, M.; Takezawa, Y. Study of high thermal conductive epoxy resins containing controlled high-order structures. *J. Appl. Polym. Sci.* **2003**, *89*, 2464–2467. [CrossRef]
14. Harada, M.; Ochi, M.; Tobita, M.; Kimura, T.; Ishigaki, T.; Shimoyama, N.; Aoki, H. Thermal-conductivity properties of liquid-crystalline epoxy resin cured under a magnetic field. *J. Polym. Sci. Part A Polym. Phys.* **2003**, *41*, 1739–1743. [CrossRef]
15. Tokushige, N.; Mihara, T.; Koide, N. Thermal properties and photo-polymerization of diepoxy monomers with mesogenic group. *Mol. Cryst. Liq. Cryst.* **2005**, *428*, 33–47. [CrossRef]
16. Song, S.; Katagi, H.; Takezawa, T. Study on high thermal conductivity of mesogenic epoxy resin with spherulite structure. *Polymer* **2012**, *53*, 4489–4492. [CrossRef]
17. Harada, M.; Hamaura, N.; Ochi, M.; Agari, Y. Thermal conductivity of liquid crystalline epoxy/BN filler composites having ordered network structure. *Compos. Part B* **2013**, *55*, 306–313. [CrossRef]
18. Li, Y.; Badrinarayanan, P.; Kessler, M. Liquid crystalline epoxy resin based on biphenyl mesogen: Thermal characterization. *Polymer* **2013**, *54*, 3017–3025. [CrossRef]
19. Giang, T.; Kim, J. Effect of backbone moiety in diglycidylether-terminated liquid crystalline epoxy on thermal conductivity of epoxy/alumina composite. *J. Ind. Eng. Chem.* **2015**, *30*, 77–84. [CrossRef]
20. Guo, H.; Zheng, H.; Gan, J.; Liang, L.; Wu, K.; Lu, M. High thermal conductivity epoxies containing substituted biphenyl mesogenic. *J. Mater. Sci. Mater. Electron.* **2016**, *27*, 2754–2759. [CrossRef]
21. Kawamoto, S.; Fujiwara, H.; Nishimura, S. Hydrogen characteristics and ordered structure of mono-mesogen type liquid-crystalline epoxy polymer. *Int. J. Hydrog. Energy* **2016**, *41*, 7500–7510. [CrossRef]
22. Guo, H.; Lu, M.; Liang, L.; Wu, K.; Ma, D.; Xue, W. Liquid crystalline epoxies with lateral substituents showing a low dielectric constant and high thermal conductivity. *J. Electron. Mater.* **2017**, *46*, 982–991. [CrossRef]
23. Kim, Y.; Yeo, H.; You, N.; Jang, S.G.; Ahn, S.; Jeong, J.; Lee, S.H.; Goh, M. Highly thermal conductive resins formed from wide-temperature-range eutectic mixtures of liquid crystalline epoxies bearing diglycidyl moieties at the side positions. *Polym. Chem.* **2017**, *8*, 2806–2814. [CrossRef]
24. Tanaka, S.; Hojo, F.; Takezawa, Y.; Kanie, K.; Muramatsu, A. Layer structure formation of mesogenic liquid crystalline epoxy resin during curing reactions: A reactive coarse-grained molecular dynamics study. *Polym.-Plast. Technol. Eng.* **2018**, *57*, 269–275. [CrossRef]
25. Tanaka, S.; Hojo, F.; Takezawa, Y.; Kanie, K.; Muramatsu, A. Homeotropically aligned monodomain-like smectic-A structure in liquid crystalline epoxy films: Analysis of the local ordering structure by microbeam small-angle X-ray scattering. *ACS Omega* **2018**, *3*, 3562–3570. [CrossRef] [PubMed]
26. Jeong, I.; Kim, C.B.; Kang, D.; Jeong, K.; Jang, S.G.; You, N.; Ahn, S.; Lee, D.; Goh, M. Liquid crystalline epoxy resin with improved thermal conductivity by intermolecular dipole–dipole interactions. *J. Polym. Sci. Part A Polym. Chem.* **2019**, *57*, 708–715. [CrossRef]
27. Ota, S.; Yamaguchi, K.; Harada, M. Phase structure and thermal conductivity of liquid crystalline epoxy resins cured with the binary mixed curing agents. *J. Netw. Polym. Jpn.* **2019**, *40*, 278–286.
28. Lin, Z.; Cong, Y.; Zhang, B.; Huang, H. Synthesis and characterization of a novel Y-shaped liquid crystalline epoxy and its effect on isotropic epoxy resin. *Liq. Cryst.* **2019**, *46*, 1467–1477. [CrossRef]
29. Shen, W.; Cao, Y.; Zhang, C.; Yuan, X. Network morphology and electro-optical characterization of epoxy based polymer stabilized liquid crystals. *Liq. Cryst.* **2020**, *47*, 481–488. [CrossRef]
30. Ota, S.; Harada, H. Filler surface adsorption of mesogenic epoxy for LC Epoxy/MgO composites with high thermal conductivity. *Compos. Part C Open Access* **2021**, *4*, 100087. [CrossRef]

31. Ota, S.; Harada, M. Thermal conductivity enhancement of liquid crystalline epoxy/MgO composites by formation of highly ordered network structure. *J. Appl. Polym. Sci.* **2021**, *138*, 50367. [CrossRef]
32. Harada, M.; Kawasaki, Y. High toughness and thermal conductivity of thermosets from liquid crystalline epoxy with low melting point. *J. Appl. Polym. Sci.* **2022**, *139*, 52391. [CrossRef]
33. Zhong, X.; Yang, X.; Ruan, K.; Zhang, J.; Zhang, H.; Gu, J. Discotic Liquid Crystal Epoxy Resins Integrating Intrinsic High Thermal Conductivity and Intrinsic Flame Retardancy. *Macromol. Rapid Commun.* **2022**, *43*, 2100580. [CrossRef]
34. Hossain, M.; Olamilekan, A.; Jeong, H. Diacetylene-containing dual-functional liquid crystal epoxy resin: Strategic phase control for topochemical polymerization of diacetylenes and thermal conductivity enhancement. *Macromolecules* **2022**, *55*, 11, 4402–4410. [CrossRef]
35. Wang, M.; Yu, Y.; Wu, X.; Li, S. Synthesis and properties of cycloaliphatic epoxy resins containing imide and diphenyl sulfone. *Polymer* **2004**, *45*, 1253–1259. [CrossRef]
36. Harada, M.; Yamaguchi, K. Fracture toughness of highly ordered liquid crystalline epoxy thermosets achieved by optimized composition of curing agents. *J. Appl. Polym. Sci.* **2021**, *138*, 50593. [CrossRef]
37. Van Krevelen, D.; Te Nijenhuis, K. *Properties of Polymers*, 4th ed.; Elsevier: Oxford, UK, 2009.

Disclaimer/Publisher's Note: The statements, opinions and data contained in all publications are solely those of the individual author(s) and contributor(s) and not of MDPI and/or the editor(s). MDPI and/or the editor(s) disclaim responsibility for any injury to people or property resulting from any ideas, methods, instructions or products referred to in the content.

Angular-Dependent Back-Reflection of Chiral-Nematic Liquid Crystal Microparticles as Multifunctional Optical Elements

Tomoki Shigeyama, Kohsuke Matsumoto, Kyohei Hisano *,† and Osamu Tsutsumi *

Department of Applied Chemistry, Ritsumeikan University, 1-1-1 Nojihigashi, Kusatsu 525-8577, Japan; sc0060kf@ed.ritsumei.ac.jp (T.S.)
* Correspondence: hisano@res.titech.ac.jp (K.H.); tsutsumi@sk.ritsumei.ac.jp (O.T.)
† Present address: Laboratory for Chemistry and Life Science, Institute of Innovative Research, Tokyo Institute of Technology, 4259 Nagatsuta, Midori-ku, Yokohama 226-8501, Japan.

Abstract: The development of multifunctional optical elements capable of controlling polarization, wavelength, and propagation direction is pivotal for the miniaturization of optical devices. However, designing the spatial distribution of the refractive index for the fabrication of such elements remains challenging. This study demonstrates the spectroscopic function of microparticles composed of chiral-nematic liquid crystals (N* LC), which inherently selectively reflect circularly polarized light. The measurement of the reflection spectra with fiber probes revealed angular-dependent back-reflection of the single layer of the N* LC particles. These results indicate that our N* LC microparticles possess multiple optical functions, enabling the separation of incident light polarization and wavelength within a single material. This suggests broad applications of N* LC particles as compact optical elements.

Keywords: chiral-nematic liquid crystals; polymer particles; selective reflection; optical elements

1. Introduction

The control of light properties, including polarization, wavelength, and propagation, is a fundamental technology with applications across various fields [1–4]. This technology is used in spectroscopy [5], sensing [6], imaging [7], three-dimensional scanning [8,9], and photovoltaics [10], which play pivotal roles in daily life and science and technology. Recently, a demand for the miniaturization of these optical devices has emerged. One approach to achieving miniaturization involves reducing the number of components. Typically, optical devices with specific functionalities are assembled using various optical elements, such as diffraction gratings, prisms, filters, polarizers, lenses, and mirrors [1]. Rational design of the spatial distribution of the refractive index, such as liquid crystal holograms, is a promising approach for realizing multifunctional optical elements. However, developing multifunctional materials for microscale components remains a significant challenge. Consequently, innovative designs for optical elements are imperative to expedite the further miniaturization of optical devices.

One potential material with such multifunctionality is a chiral-nematic liquid crystal (N* LC), which exhibits wavelength-selective reflection of circularly polarized light (CPL) [11–17]. N* LC is a liquid–crystalline phase that emerges upon the introduction of chiral molecules into nematic liquid crystals, spontaneously forming a helical molecular alignment. Owing to its helical structure and refractive index anisotropy, this phase exhibits a periodic distribution of the refractive index and demonstrates selective reflection, following Bragg's law. The reflection wavelength λ_{ref} depends on their helical pitch P in which the molecules rotate 360°, as per the following equation:

$$\lambda_{\text{ref}} = n_{\text{N* LC}} \cdot P \sin\theta,$$

where $n_{N^* LC}$ is the average refractive index of the N* LC used and θ is the angle between the incident light and surface plane. By manipulating the molar ratio of the chiral molecules, we can modulate the helical pitch (P) of the N* LC, allowing us to attain N* LC with a specified reflection wavelength. Moreover, the N* LC exhibited selective reflection of CPL with the same handedness as its helical structure [18]. The dual functionality of N* LC, which enables simultaneous control over the circular polarization and wavelength of reflected light, has generated considerable interest in its application as a multifunctional optical element.

In addition to these functionalities, several studies have successfully controlled the light propagation direction [11–15]. Recently, Ozaki et al. proposed an innovative route to diffuse or focus reflected light by manipulating the phase of the helical alignment along the helical axis within the film [11]. This new capability to control the light direction holds a significant potential for advancing the application of N* LC as a multifunctional holographic device. However, the propagation direction of light depends on the incidence angle. This is because of the film-type N* LC with a unidirectionally aligned helical axis. Furthermore, these functionalities were exclusively observed under monochromatic light. Achieving back-reflection capable of changing the reflection wavelength based on the incidence angle remains challenging for N* LC materials.

We recently fabricated N* LC particles through the dispersion polymerization of monomers [19,20]. This technology enables the production of monodisperse microsized polymer particles in a single step [21,22]. The resulting N* LC particles exhibited vivid reflection colors that varied with the molar ratio of the chiral monomer. These polymer particles can serve as reflective materials with superior environmental stability compared with low-molecular-weight LC droplets [23–26]. In a previous study, molecular alignments revealed the presence of radial helical axis alignments [19,20]. Because the helical axis alignment is not unidirectional, the incidence angle against the helical axis varies according to the illuminated area of the N* LC particles [27]. We anticipated that this heterogeneous reflection behavior would induce angular-dependent back-reflection. This is because the reflection wavelength and angle were altered based on the orientation of the incident beam.

Herein, we conducted experimental investigations into the angular-dependent reflective properties of the N* LC particles to assess their potential as versatile optical elements. We observed variations in the back-reflection wavelength corresponding to changes in the incidence angle. These multifunctional N* LC particles hold promise for applications as ultracompact optical elements at the micrometer scale. The N* LC particles exhibit reflective properties at specific wavelengths and CPL selectivity. In addition, these particles can produce diverse reflection colors, enabling the straightforward creation of colorant materials with a range of reflection modes [19,20]. Thus, our microsized N* LC particles represent a substantial advancement in developing multifunctional optical elements, significantly contributing to the miniaturization of optical devices.

2. Materials and Methods

2.1. Preparation of N* LC Particles Using Dispersion Polymerization

The N* LC particles were synthesized following the procedure outlined in our previous report [19,28]. The molecular structures of the monomers and dispersion stabilizer (polyvinylpyrrolidone (PVP)) are shown in Figure 1. The base liquid crystal monomer (LCM) was generously provided by Osaka Organic Chemical Industry Ltd (Osaka, Japan). and was purified by recrystallization from methanol before use. The chiral monomer (CM) was synthesized and purified according to previously reported methods [28].

The monomers, LCM and CM, along with PVP and polymerization initiator (2,2'-azobis(isobutyronitrile), AIBN), were dissolved in a mixed solvent of N,N-dimethylformamide (DMF) and methanol in a 30 mL Schlenk flask (U-1013, Sugiyama-Gen Co., Ltd., Tokyo, Japan). The composition of the polymerization mixture is listed in Table S1. The solution was subjected to several freeze–pump–thaw cycles to remove dissolved oxygen, backfilled with Ar gas, and then stirred at 55 °C for 20 h. The resulting N* LC particles were collected by filtration through a membrane filter (T080A025A, ADVANTEC, Tokyo, Japan) with a pore size of 0.8 μm to obtain the polymer particles.

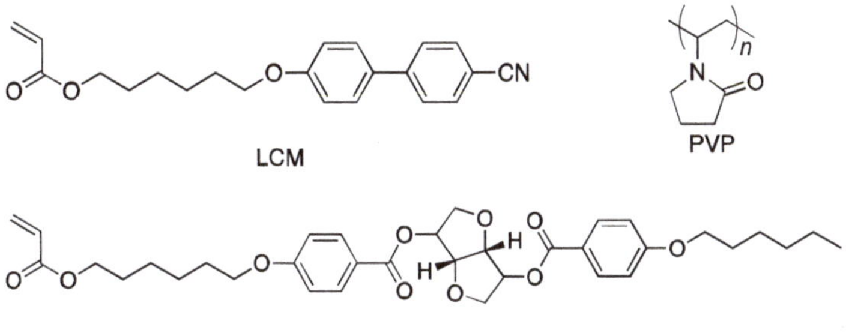

Figure 1. Molecular structures of liquid crystal monomer (LCM), chiral monomer (CM), and dispersion stabilizer (PVP).

2.2. Characterization of N* LC Particles

Size-exclusion chromatography (SEC) was performed using LC-20AD (Shimadzu, Kyoto, Japan) equipped with a KF805 column (Shodex, Tokyo, Japan) and UV-vis detector (254 nm) at 40 °C, using tetrahydrofuran (THF) as an eluent at a flow rate of 1.0 mL min^{-1}. Molecular weights were calibrated using polystyrene standards. ^{1}H NMR spectra were recorded on a JEOL ECS-400 spectrometer (400 MHz) in CDCl$_3$ for the monomers and CD$_2$Cl$_2$ for the polymer. Chemical shifts were reported in parts per million (ppm) using the residual protons in the NMR solvent. The thermodynamic properties were determined using differential scanning calorimetry (DSC, SII X-DSC7000) at heating and cooling rates of 10 °C min^{-1}. Representative results of the P2 measurements are shown in Figures S1, S2 and 2.

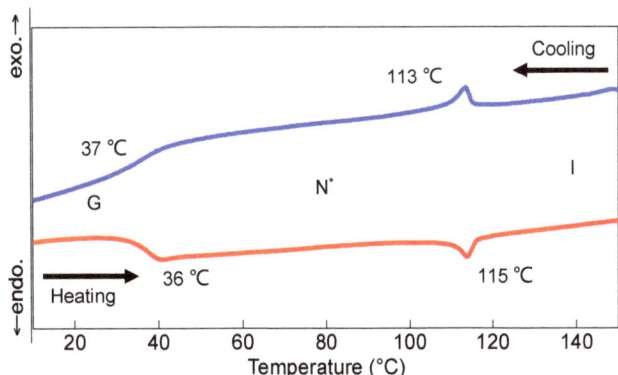

Figure 2. DSC thermograms of P2 (second scan cycle). Scanning rate was 10 °C min^{-1}. Abbreviations: G, glassy; N*, chiral-nematic; I, isotropic.

2.3. Evaluation of Reflection Functionalities

Single layer particles were fabricated on the carbon tape as follows. A suspension of the N* LC polymer particles (1.0 mg mL^{-1}, 0.2 mL) in water was dropcasted onto a pre-cleaned glass substrate (15 mm × 13 mm) and slowly evaporated at 5 °C overnight. Subsequently, a carbon tape was adhered to this substrate, and a single layer of N* LC particles was transferred onto the carbon tape.

Reflection spectra were measured using a diode array spectrometer (BLUE-WaveUVN, StellarNet, Tampa, FL, USA) equipped with two types of fiber probes. A coaxial fiber probe

(R600, StellarNet, Tampa, FL, USA) was employed to measure the normal reflection. To investigate the incidence angle dependence, two uniaxial fiber probes (F400, StellarNet, Tampa, FL, USA) were used. In this study, the incidence angle (θ) was defined as the angle between the substrate surface plane and the incident light, which was controlled by the fiber probe connected to the light source. The fiber probe connected to the detector was fixed perpendicular to the substrate surface. A tungsten lamp (SL-1, StellarNet, Tampa, FL, USA) served as the incident light source for unpolarized white light. To evaluate the differences in wavelengths, each spectrum was normalized by dividing the original data by the corresponding intensity value at the maximum reflection wavelength in the measured ranges.

3. Results and Discussion

3.1. Synthesis and Characterization of N* LC Polymer Particles

We successfully obtained monodisperse polymer particles with an average diameter (d) of ~2.5 µm and a standard deviation of 0.1 µm at ~50% conversion (Table 1). SEC analysis revealed a number-average molecular weight (M_n) of 15,000 Da and a polydispersity index of 2.6 (Table 1). The copolymer composition was determined using ^{1}H NMR spectroscopy, which confirmed that the desired copolymer with the same composition as that of the monomer mixture for polymerization was achieved (Table 1 and Figure S2). DSC measurement indicates that the synthesized polymers had a glass transition temperature (T_g) at 36 °C and showed a liquid–crystalline phase up to 115 °C in the heating process (Figure 2). These results demonstrate that our N* LC particles were solid at room temperature, wherein the helical alignment of the mesogens in the N* LC phase was fixed within the solid particles.

Table 1. General properties of the N* LC polymer particles.

Particle	Molar ratio of CM (mol%)	d (µm)	M_n (M_w/M_n)	T_g (°C)
P1	2.4	2.5	15,000 (2.6)	36
P2	3.8	2.6	15,000 (2.6)	37

Abbreviations: d, average diameter; M_n, number-average molecular weight; M_w, weight-average molecular weight; T_g, glass transition temperature during heating.

Microparticles with red (P1) and blue (P2) reflection colors were prepared by controlling the molar ratio of the CM (Figure 3a). Observations through CPL filters revealed that right-handed CPL was reflected by N* LC particles (Figure 3b) [18]. As in our previous report, we considered that the helical axes aligned radially within the particles, as evidenced by the Maltese cross-pattern observed using polarized optical microscopy (POM, Figure 3c) [19]. The idealized molecular and helical axis alignments are schematically illustrated in Figures 3d and 3e, respectively.

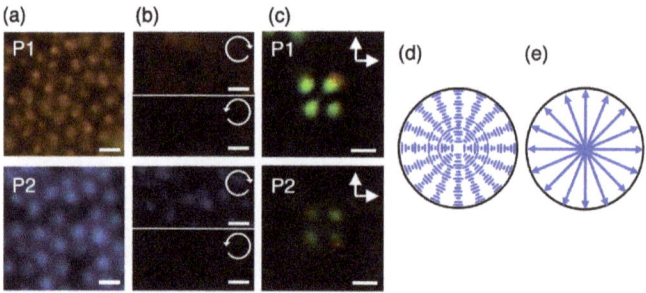

Figure 3. (**a,b**) Epi-illuminated micrographs of P1 (**upper**) and P2 (**lower**) without (**a**) and with CPL filter (**b**). In (**b**), white circle-arrows show the handedness of light transmitted through the CPL filter.

Scale bars represent 5 µm. (**c**) POM images of P1 and P2 dispersed in water. White arrows show the direction of polarizers. Scale bars represent 1 µm. (**d,e**) Schematics of the idealized molecular alignment (**d**) and the helical axis alignment (**e**) in the N* LC particles.

3.2. Retro-Reflection Behavior

We measured the retro-reflection spectra from a single layer of N* LC particles with an incidence angle normal to the substrate surface using a coaxial optical fiber probe, as shown in Figure 4a. In the obtained spectra, prominent bands were observed at 760 nm for P1 and 480 nm for P2, along with several smaller bands at baseline (Figure 4b). The wavelength of the major reflection peak varied with the molar ratio of the chiral agent, indicating its association with the selective reflection from the particles. The bandwidth of P1 was wider than that of P2 because the bandwidth of the reflection from the N* LC was proportional to the helical pitch. The reflection bandwidth $\Delta\lambda_{ref}$ is determined by the following equation when light is normally incident to the helical axis: $\Delta\lambda_{ref} = \Delta n_{N^* LC} \cdot P$, where $\Delta n_{N^* LC}$ is a birefringence of mesogen. Assuming that the birefringence is constant in the wavelength range of visible light (400–800 nm), the $\Delta\lambda_{ref}$ depends on the value of P. Therefore, P1 with the longer helical pitch, i.e., longer reflection wavelength, should possess a wider reflection bandwidth than P2 with the shorter helical pitch. Thus, nearly identical reflection wavelengths were previously reported in reflection spectroscopy using an integration sphere (Figure S3) [19,20]. However, a detailed comparison between the spectra obtained with the fiber probe and integration sphere revealed that the reflection peak detected with the fiber probe was narrower, with a slight red shift. We attributed this discrepancy to the characteristics of the optical setup employed for the measurements.

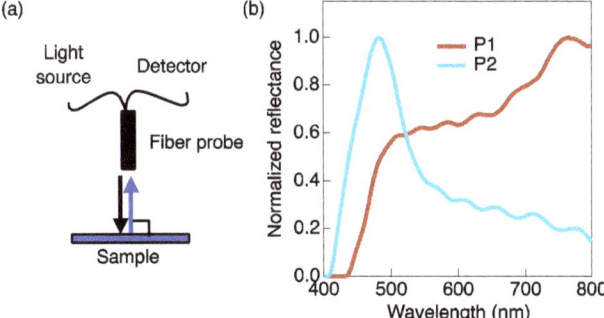

Figure 4. (**a**) Optical setup for measuring reflection spectra using the coaxial fiber probe. Black and blue arrows indicate the incident and reflection light, respectively. (**b**) Reflection spectra of the single-layered N* LC polymer particles, P1 and P2.

In measurements using the coaxial fiber probe, only the light reflected perpendicular to the tangent of the particle, essentially parallel to the helical axis, can be detected. This indicated that only specular reflections from the particles were captured, resulting in the observation of narrow reflection bands. Conversely, omnidirectional reflections from the particles at all solid angles were amalgamated in the measurements using the integrating sphere. Consequently, the shorter reflection wavelengths stemming from the tilted helical axis caused the reflection bands to broaden and shift toward the shorter end of the spectrum. These findings strongly imply that the wavelength of light reflected by the particles is contingent on the direction of the incident light.

The small bands observed at the baseline of the reflection spectra are interpreted as interference fringes resulting from the densely packed single microparticle layer [29]. The peak wavelengths of the neighboring interference fringes (λ_i and λ_{i+1}, where i is a consecutive number of fringes) allow us to calculate the thickness of the layer, denoted as d', which can be theoretically estimated as follows:

$$d' = \frac{\lambda_i \cdot \lambda_{i+1}}{2n_{\text{eff}}(\lambda_i - \lambda_{i+1})},$$

where n_{eff} is the effective refractive index of the packed structure, given by the following equation [29]:

$$n_{\text{eff}}^2 = \psi\, n_{\text{N}^* \text{LC}}^2 + (1 - \psi)\, n_{\text{air}}^2,$$

where ψ represents the packing density of the particles and n_{air} is the refractive index of air. When the particles are arranged in a single layer with a two-dimensional closest packing structure, ψ is 0.61. $n_{\text{N}^* \text{LC}}$ can be the same as that of a low-molecular-weight liquid crystal with a chemical structure similar to monomer LCM, namely, 4′-pentyl-4-biphenylcarbonitrile (n = 1.6) [30]. This implies that the n_{eff} can be estimated as 1.4. Using this value and considering the periods of neighboring interference fringes in the reflection spectra of P2, d' was calculated to be 2.4 µm. This value closely corresponds to d, supporting our hypothesis that the small peaks observed at the baseline are indeed interference fringes from the microparticle layer. Based on these results, we confirmed that the microparticles formed a single layer with the closest packing arrangement on the substrate.

3.3. Optical Functionalities under Different Incidence Angles

To investigate the dependence of the incidence angle on the wavelength of reflection normal to the substrate, we conducted reflection spectra measurements using two uniaxial fiber probes while varying the angle θ, as shown in Figure 5a. θ is defined as the angle between the substrate and incident light. The fiber probe connected to the detector was fixed perpendicular to the substrate. In the generally planarly aligned N* LC film, the reflection angle corresponds to the incidence angle owing to specular reflection following Bragg's law. This implies that the back-reflection is not observed; thus, for the optical setup in Figure 5a with discordance between incident and reflection angles against the sample plane, the reflection peak should be extremely small or not be recorded, except at θ = 90°. However, as shown in Figure 5b, a reflection band was observed at all angles, even when the incidence angle differed from the reflection angle. Furthermore, the reflection wavelength shifted towards shorter wavelengths as θ decreased. According to Bragg's law, the reflection wavelength is determined by the angle between the incident light and helical axis [27]. This geometric relationship is shown in Figure 5c. We anticipated that the radial alignment of the helical axis would result in a distinctive reflection pattern, where the incident and reflection angles would match at the tangent of the particle surface. Following this expectation, the reflection wavelength is determined by the angle α between the incident light and particle tangent (Figure 5c). In this illustration, the angle between the incident light and the reflected light is 90° $-$ θ. The angle between the normal to helical axis and the reflected light is also α. Therefore, the relationship between θ and α can be expressed as α + 90° $-$ θ + α = 180°. The relationship between α and θ is best described as follows:

$$\alpha = 45° + \theta/2.$$

From this relationship and Bragg's law, the reflection wavelength at each θ, λ_θ, was estimated using the following equation [31]:

$$\lambda_\theta = \lambda_{90°} \sin \alpha.$$

The estimated and experimental values of λ_θ are plotted in Figure 5d, demonstrating a close match between them. These findings indicate that the back-reflection wavelength of N* LC particles varies depending on the incidence angle. Given that each particle exhibits this angular-dependent back-reflection, it can be used as an ultracompact optical element.

In addition, the N* LC particles showed right-handed CPL-selective reflections (Figure 3b). Consequently, our N* LC particles can control the circular polarization, wavelength, and direction of the incident light, even with a single element. Therefore, they can be used as multifunctional optical elements, contributing to the miniaturization of optical devices, including MEMS and microsensors.

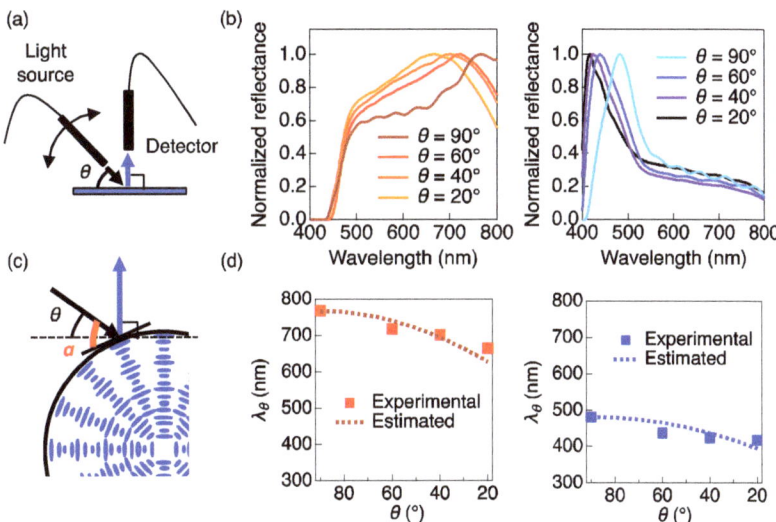

Figure 5. (a) Optical setup for measuring reflection spectra with two uniaxial optical fiber probes. Black and blue arrows indicate the incident light and reflection light, respectively. The θ (°), an angle between the incident light and substrate, varied while the fiber probe of the detector side was fixed perpendicular to the substrate. (b) Reflection spectra of the N* LC polymer particles P1 (**left**) and P2 (**right**) when θ changed. (c) Schematic of the reflection manner of the N* LC polymer particles. α (°) is an angle between the incident light and particle tangent. (d) θ dependence of the peak top of reflection spectra in (b). The dashed line indicates the peak top estimated from the equation $\lambda_\theta = \lambda_{90°} \sin(45° + \theta/2)$.

4. Conclusions

In this study, we synthesized N* LC microparticles via dispersion polymerization and characterized their unique reflective functionalities. We measured the CPL reflection wavelength, which depended on the incidence angle, revealing the CPL-selective angular-dependent back-reflection of the N* LC particles. This underscores their potential as multifunctional optical elements for controlling the circular polarization, wavelength, and direction of incident light. Notably, compared to the low-molecular-weight N* LC droplets developed in recent years, our N* LC particles exhibited stability in the solid state [23,31–34], augmenting their utility as ultrasmall optical elements. Furthermore, liquid crystal materials have the potential to expand functionalities through stimuli responsiveness [35,36]. As such, our N* LC particles represent a significant advancement in the field of versatile optical elements and offer a new avenue for designing compact optical devices.

5. Patents

The methodology for generating the N* LC particles outlined in this paper was filed as the following patents by Ritsumeikan University: JP 2020-139135A (2020) and PCT/JP2022/019448 (2022).

Supplementary Materials: The following supporting information can be downloaded from https://www.mdpi.com/article/10.3390/cryst13121660/s1: Figure S1: Size-exclusion chromatogram of P2.; Figure S2: ^{1}H NMR spectra of the monomers and P2.; Figure S3: Reflection spectra of the N* LC polymer particles of P1 and P2 were measured with an integration sphere.; Table S1: Polymerization composition for synthesis of the N* LC particles.

Author Contributions: Conceptualization, K.H. and O.T.; methodology, T.S.; investigation, T.S., K.M. and K.H.; writing—original draft preparation, T.S., K.M., K.H. and O.T.; supervision, K.H. and O.T.; project administration, K.H. and O.T.; funding acquisition, T.S., K.H. and O.T. All authors have read and agreed to the published version of the manuscript.

Funding: This research was funded by the Japan Science and Technology Agency (JST) A-STEP (JPMJTR22T1, for O.T.), the Toshiaki Ogasawara Memorial Foundation (for O.T.), the Japan Society for the Promotion of Science (JSPS) KAKENHI (22K14737, for K.H.), JST SPRING (JPMJSP2101, for T.S.), and the Cooperative Research Program of the Network Joint Research Center for Materials and Devices.

Data Availability Statement: The data presented in this study are available in this article and supplementary materials.

Acknowledgments: We thank Osaka Organic Chemical Industry, Ltd. for generously providing the base liquid crystal monomer (LCM).

Conflicts of Interest: The authors declare no conflict of interest.

References

1. Li, L.; Bryant, D.; Bos, P.J. Broadband Liquid crystal lens with concentric electrodes and inter-electrode resistors. *Liq. Cryst. Rev.* **2014**, *2*, 130–154. [CrossRef]
2. Ishiguro, M.; Sato, D.; Shishido, A.; Ikeda, T. Bragg-Type Polarization Gratings Formed in Thick Polymer Films Containing Azobenzene and Tolane Moieties. *Langmuir* **2007**, *23*, 332–338. [CrossRef]
3. Provenzano, C.; Pagliusi, P.; Cipparrone, G. Highly Efficient Liquid Crystal Based Diffraction Grating Induced by Polarization Holograms at the Aligning Surfaces. *Appl. Phys. Lett.* **2006**, *89*, 121105. [CrossRef]
4. Foelen, Y.; Schenning, A.P.H.J. Optical Indicators Based on Structural Colored Polymers. *Adv. Sci.* **2022**, *9*, 2200399. [CrossRef] [PubMed]
5. Eichert, D.; Gregoratti, L.; Kaulich, B.; Marcello, A.; Melpignano, P.; Quaroni, L.; Kiskinova, M. Imaging with Spectroscopic Micro-Analysis Using Synchrotron Radiation. *Anal. Bioanal. Chem.* **2007**, *389*, 1121–1132. [CrossRef] [PubMed]
6. Ferreira, M.F.S.; Castro-Camus, E.; Ottaway, D.J.; López-Higuera, J.M.; Feng, X.; Jin, W.; Jeong, Y.; Picqué, N.; Tong, L.; Reinhard, B.M. Roadmap on Optical Sensors. *J. Opt.* **2017**, *19*, 083001. [CrossRef]
7. Zhu, X.; Xia, Y.; Wang, X.; Si, K.; Gong, W. Optical Brain Imaging: A Powerful Tool for Neuroscience. *Neurosci. Bull.* **2017**, *33*, 95–102. [CrossRef]
8. Talbot, B.; Astrup, R. A Review of Sensors, Sensor-Platforms and Methods Used in 3D Modelling of Soil Displacement after Timber Harvesting. *Croat. J. For. Eng.* **2021**, *42*, 149–164. [CrossRef]
9. Raj, T.; Hashim, F.H.; Huddin, A.B.; Ibrahim, M.F.; Hussain, A. A Survey on LiDAR Scanning Mechanisms. *Electronics* **2020**, *9*, 741. [CrossRef]
10. Shanks, K.; Senthilarasu, S.; Mallick, T.K. Optics for Concentrating Photovoltaics: Trends, Limits and Opportunities for Materials and Design. *Renew. Sustain. Energy Rev.* **2016**, *60*, 394–407. [CrossRef]
11. Kobashi, J.; Yoshida, H.; Ozaki, M. Planar Optics with Patterned Chiral Liquid Crystals. *Nat. Photonics* **2016**, *10*, 389–392. [CrossRef]
12. Yue, Y.; Kurokawa, T.; Haque, M.A.; Nakajima, T.; Nonoyama, T.; Li, X.; Kajiwara, I.; Gong, J.P. Mechano-Actuated Ultrafast Full-Colour Switching in Layered Photonic Hydrogels. *Nat. Commun.* **2014**, *5*, 4659. [CrossRef] [PubMed]
13. Kim, D.Y.; Nah, C.; Kang, S.W.; Lee, S.H.; Lee, K.M.; White, T.J.; Jeong, K.U. Free-Standing and Circular-Polarizing Chirophotonic Crystal Reflectors: Photopolymerization of Helical Nanostructures. *ACS Nano* **2016**, *10*, 9570–9576. [CrossRef]
14. Chen, P.; Ma, L.L.; Duan, W.; Chen, J.; Ge, S.J.; Zhu, Z.H.; Tang, M.J.; Xu, R.; Gao, W.; Li, T. Digitalizing Self-Assembled Chiral Superstructures for Optical Vortex Processing. *Adv. Mater.* **2018**, *30*, 1705865. [CrossRef]
15. Hisano, K.; Kimura, S.; Ku, K.; Shigeyama, T.; Akamatsu, N.; Shishido, A.; Tsutsumi, O. Mechano-Optical Sensors Fabricated with Multilayered Liquid Crystal Elastomers Exhibiting Tunable Deformation Recovery. *Adv. Funct. Mater.* **2021**, *31*, 2104702. [CrossRef]
16. White, T.J.; McConney, M.E.; Bunning, T.J. Dynamic Color in Stimuli-Responsive Cholesteric Liquid Crystals. *J. Mater. Chem.* **2010**, *20*, 9832–9847. [CrossRef]
17. Mitov, M. Cholesteric Liquid Crystals with a Broad Light Reflection Band. *Adv. Mater.* **2012**, *24*, 6260–6276. [CrossRef]

18. Shin, S.; Park, M.; Ku Cho, J.; Char, J.; Gong, M.; Jeong, K.U. Tuning Helical Twisting Power of Isosorbide-Based Chiral Dopants by Chemical Modifications. *Mol. Cryst. Liq. Cryst.* **2011**, *534*, 19–31. [CrossRef]
19. Shigeyama, T.; Hisano, K.; Tsutsumi, O. Control of Helical-Axis Orientation of Chiral Liquid Crystals in Monodispersed Polymer Particles. *Proc. SPIE* **2021**, *11807*, 118070F. [CrossRef]
20. Shigeyama, T.; Hisano, K.; Matsumoto, K.; Tsutsumi, O. Tunable Reflection through Size Polydispersity of Chiral-Nematic Liquid Crystal Polymer Particles. *Molecules* **2023**, *28*, 7779. [CrossRef]
21. Kawaguchi, S.; Ito, K. Dispersion Polymerization. *Adv. Polym. Sci.* **2005**, *175*, 299–328. [CrossRef]
22. Paine, A.J.; Luymes, W.; McNulty, J. Dispersion Polymerization of Styrene in Polar Solvents. 6. Influence of Reaction Parameters on Particle Size and Molecular Weight in Poly(*N*-Vinylpyrrolidone)-Stabilized Reactions. *Macromolecules* **1990**, *23*, 3104–3109. [CrossRef]
23. Chen, H.Q.; Wang, X.Y.; Bisoyi, H.K.; Chen, L.J.; Li, Q. Liquid Crystals in Curved Confined Geometries: Microfluidics Bring New Capabilities for Photonic Applications and Beyond. *Langmuir* **2021**, *37*, 3789–3807. [CrossRef] [PubMed]
24. Balenko, N.; Shibaev, V.; Bobrovsky, A. Mechanosensitive polymer-dispersed cholesteric liquid crystal composites based on various polymer matrices. *Polymer* **2023**, *281*, 126119. [CrossRef]
25. Pan, Y.; Xie, S.; Wang, H.; Huang, L.; Shen, S.; Deng, Y.; Ma, Q.; Liu, Z.; Zhang, M.; Jin, M.; et al. Microfluidic Construction of Responsive Photonic Microcapsules of Cholesteric Liquid Crystal for Colorimetric Temperature Microsensors. *Adv. Opt. Mater.* **2023**, *11*, 2202141. [CrossRef]
26. Froyen, A.A.F.; Debije, M.G.; Schenning, A.P.H.J. Polymer Dispersed Cholesteric Liquid Crystal Mixtures for Optical Time–Temperature Integrators. *Adv. Opt. Mater.* **2022**, *10*, 2201648. [CrossRef]
27. Agha, H.; Zhang, Y.S.; Geng, Y.; Lagerwall, J.P.F. Pixelating Structural Color with Cholesteric Spherical Reflectors. *Adv. Photonics Res.* **2023**, *4*, 2200363. [CrossRef]
28. Ku, K.; Hisano, K.; Kimura, S.; Shigeyama, T.; Akamatsu, N.; Shishido, A.; Tsutsumi, O. Environmentally Stable Chiral-nematic Liquid-crystal Elastomers with Mechano-optical Properties. *Appl. Sci.* **2021**, *11*, 5037. [CrossRef]
29. Jiang, P.; Bertone, J.F.; Hwang, K.S.; Colvin, V.L. Single-Crystal Colloidal Multilayers of Controlled Thickness. *Chem. Mater.* **1999**, *11*, 2132–2140. [CrossRef]
30. Mamuk, A.E.; Nesrullajev, A.; Mukherjee, P.K. Refractive and Birefringent Properties of 4-Alkyl-4′-Oxycyanobiphenyls at Direct and Reverse Phase Transitions. *Mol. Cryst. Liq. Cryst.* **2017**, *648*, 168–181. [CrossRef]
31. Noh, J.H.; Liang, H.-L.; Olenikcd, I.D.; Lagerwall, J.P.F. Tuneable multicoloured patterns from photonic cross-communication between cholesteric liquid crystal droplets. *J. Mater. Chem. C* **2014**, *2*, 806–810. [CrossRef]
32. He, J.; Liu, S.; Gao, G.; Sakai, M.; Hara, M.; Nakamura, Y.; Kishida, H.; Seki, T.; Takeoka, Y. Particle Size Controlled Chiral Structural Color of Monodisperse Cholesteric Liquid Crystals Particles. *Adv. Opt. Mater.* **2023**, *11*, 2300296. [CrossRef]
33. Humar, M.; Muševič, I.; Sullivan, K.G.; Hall, D.G. 3D Microlasers from Self-Assembled Cholesteric Liquid-Crystal Microdroplets. *Opt. Express* **2010**, *18*, 26995–27003. [CrossRef] [PubMed]
34. Fan, J.; Li, Y.; Bisoyi, H.K.; Zola, R.S.; Yang, D.; Bunning, T.J.; Weitz, D.A.; Li, Q. Light-Directing Omnidirectional Circularly Polarized Reflection from Liquid-Crystal Droplets. *Angew. Chem. Int. Ed.* **2015**, *54*, 2160–2164. [CrossRef] [PubMed]
35. Belmonte, A.; Pilz da Cunha, M.; Nickmans, K.; Schenning, A.P.H.J. Brush-Paintable, Temperature and Light Responsive Triple Shape-Memory Photonic Coatings Based on Micrometer-Sized Cholesteric Liquid Crystal Polymer Particles. *Adv. Opt. Mater.* **2020**, *8*, 2000054. [CrossRef]
36. Belmonte, A.; Ussembayev, Y.Y.; Bus, T.; Nys, I.; Neyts, K.; Schenning, A.P.H.J. Dual Light and Temperature Responsive Micrometer-Sized Structural Color Actuators. *Small* **2020**, *16*, 1905219. [CrossRef]

Disclaimer/Publisher's Note: The statements, opinions and data contained in all publications are solely those of the individual author(s) and contributor(s) and not of MDPI and/or the editor(s). MDPI and/or the editor(s) disclaim responsibility for any injury to people or property resulting from any ideas, methods, instructions or products referred to in the content.

Article

Eliminating Ambiguities in Electrical Measurements of Advanced Liquid Crystal Materials

Oleksandr V. Kovalchuk [1,2], Tetiana M. Kovalchuk [3] and Yuriy Garbovskiy [4,*]

[1] Institute of Physics, National Academy of Sciences of Ukraine, 03680 Kyiv, Ukraine; akoval@knutd.com.ua
[2] Department of Applied Physics and Higher Mathematics, Kyiv National University of Technologies and Design, 01011 Kyiv, Ukraine
[3] V. Lashkaryov Institute of Semiconductor Physics, National Academy of Sciences of Ukraine, 03680 Kyiv, Ukraine; tatnik1412@gmail.com
[4] Department of Physics and Engineering Physics, Central Connecticut State University, New Britain, CT 06050, USA
* Correspondence: ygarbovskiy@ccsu.edu

Abstract: Existing and future display and non-display applications of thermotropic liquid crystals rely on the development of new mesogenic materials. Electrical measurements of such materials determine their suitability for a specific application. In the case of molecular liquid crystals, their direct current (DC) electrical conductivity is caused by inorganic and/or organic ions typically present in small quantities even in highly purified materials. Important information about ions in liquid crystals can be obtained by measuring their DC electrical conductivity. Available experimental reports indicate that evaluation of the DC electrical conductivity of liquid crystals is a very non-trivial task as there are many ambiguities. In this paper, we discuss how to eliminate ambiguities in electrical measurements of liquid crystals by considering interactions between ions and substrates of a liquid crystal cell. In addition, we analyze factors affecting a proper evaluation of DC electrical conductivity of advanced multifunctional materials composed of liquid crystals and nanoparticles.

Keywords: liquid crystals; ions; nanomaterials; electrical measurements; electrical conductivity; ion generation

Citation: Kovalchuk, O.V.; Kovalchuk, T.M.; Garbovskiy, Y. Eliminating Ambiguities in Electrical Measurements of Advanced Liquid Crystal Materials. *Crystals* 2023, *13*, 1093. https://doi.org/10.3390/cryst13071093

Academic Editor: Ingo Dierking

Received: 27 June 2023
Revised: 7 July 2023
Accepted: 10 July 2023
Published: 13 July 2023

Copyright: © 2023 by the authors. Licensee MDPI, Basel, Switzerland. This article is an open access article distributed under the terms and conditions of the Creative Commons Attribution (CC BY) license (https://creativecommons.org/licenses/by/4.0/).

1. Introduction

Tunable liquid crystal components can be found in numerous devices and commercial products. They include ubiquitous liquid crystal displays [1–4], electrically controlled lenses [5,6], waveguides [7,8], waveplates and filters [9,10], spatial light modulators [11], diffractive elements [12,13], privacy windows and shutters [14–17], reconfigurable meta- and plasmonic devices [18–20], intensity modulators and lasers [21–23], sensors [24], and smart devices tailored to microwave and millimeter-wave applications [25–27]. As a rule, such devices rely on the reorientation of liquid crystals under the action of applied electric fields. This reorientation can be altered by ions typically present in molecular liquid crystals in minute quantities [28–31]. In the simplest case, a standard electric field screening effect caused by ions can result in unpredictable changes in the performance of liquid crystal devices [28–31]. In addition, ions in molecular liquid crystals can result in extra power losses due to the Joule heating effect. That's why electrical measurements of any new liquid crystal material are a standard part of its material characterization [32,33]. Last but not least, the presence of ions in liquid crystal materials is important for understanding the unusual behavior of ferroelectric nematic liquid crystals [34] and polymer stabilized liquid crystals [35,36].

The importance of properly performed electrical measurements of liquid crystals was emphasized in many papers [37,38]. Standard experimental techniques include dielectric and impedance spectroscopy [39–42], transient current measurements [43,44], residual direct current (DC) voltage measurements [45,46], voltage holding ratio [45,47], and flicker minimization methods [47]. A proper evaluation of the DC electrical conductivity of liquid crystals is very critical because its value determines the suitability of mesogenic materials for a particular application [32,33]. In a typical experimental setting, the measurements of the electrical resistivity ρ_{LC} or its inverse, DC electrical conductivity $\lambda_{DC} = 1/\rho_{LC}$, are routinely performed. As a rule, the DC electrical conductivity measurements of liquid crystals are carried out using a sandwich-type cell of a single thickness. Only a few papers report the values of the DC electrical conductivity and/or concentration of ions obtained for cells of several thicknesses [38,40,48–52]. Even though the process of measuring the ionic conductivity is rather straightforward, the reported values of the ionic conductivity even of the same type of liquid crystals (as an example, consider a standard nematic 5CB) can vary significantly (up to two orders of magnitude) depending on the cell thickness and time [31,40,49]. To reduce this ambiguity, in evaluating the values of the DC electrical conductivity of molecular liquid crystals it is important to consider interactions between ions and substrates of the liquid crystal cell. Such interactions can result in the time-dependent and thickness-dependent DC electrical conductivity of molecular liquid crystals [53–56]. Because systematic (i.e., combining experiment and modelling) studies of such ionic effects are still rather rare, it is both interesting and important to study them in greater detail.

Future liquid crystal technologies rely on the development of advanced liquid crystal materials. Mixing liquid crystals with nanomaterials is considered a promising way to produce such multifunctional materials [57–61]. From both academic and industrial points of view, it is crucial to study electrical properties of such materials. Research results obtained by independent teams during the last two decades point to very non-trivial ionic effects in liquid crystals doped with nanoparticles of different origins (ferroelectric [62–70], magnetic [52,71–78], semiconductor and dielectric [79–87], carbon-based [88–98], and metal [99–106], additional references can be found in recent topical reviews [31,107–109]). If nanoparticles are added to liquid crystals, an analysis of the existing literature also reveals a great variability of the reported values of ionic conductivity of seemingly similar materials making the comparison of the reported results a very difficult task [31,107–109]. As a result, a proper evaluation of the DC electrical conductivity of molecular liquid crystals doped with nanoparticles is quite a challenging problem because its value can be substantially affected by interactions between ions, nanomaterials, and substrates of a liquid crystal cell [53].

In this paper, we present an analysis of how interactions between ions, cell substrates, and nanoparticles can affect the value of the DC electrical conductivity of molecular liquid crystals in real experimental settings. We use already reported experimental results [38,50,52,54,110] and analyze them from the same conceptual approach with a hope that the presented analysis and suggestions would be useful for experimentalists measuring electrical properties of liquid crystal materials. By discussing several experimental cases in the framework of the Langmuir adsorption model, we point to the importance of the following experimentally observed outcomes of such interactions: (i) interactions between ions and substrates of a liquid crystal cell result in time dependence of the measured value of the DC electrical conductivity; (ii) the measured value of the DC electrical conductivity also depends on the cell thickness; (iii) under certain conditions, nanomaterials in molecular liquid crystals can behave as sources of ions and exhibit behavior similar to that of weak electrolytes. We also show that whenever nanoparticles are used to improve the functionality of liquid crystals, a proper electrical characterization of such advanced materials should consider the combined effect of interactions between ions, substrates, and nanoparticles. We also describe how to decouple these processes in a typical experimental setting. In addition, some suggestions to improve existing experimental procedures for the evaluation of the DC electrical conductivity of liquid crystal materials are provided.

2. Materials and Methods

The DC electrical conductivity was measured by applying a dielectric spectroscopy method [38,42,111]. Sandwich-type cells of known thickness were filled with nematic liquid crystals 5CB [54,110], MJ961180 [38,112], and 6CB [52]. Nematic liquid crystals 5CB were used to analyze the time-dependent electrical conductivity. The dependence of the DC electrical conductivity on the cell thickness was illustrated using nematic liquid crystals MJ961180, and the combined effects of the cell thickness and iron-oxide nanoparticles was studied in nematic liquid crystals 6CB. The cell thickness was controlled by means of spacers. Indium–tin oxide substrates were covered with polyimide thin films to impose either planar (5CB, MJ961180) or homeotropic (6CB) boundary conditions. Additional experimental details can be found in papers [38,52,110].

The DC electrical conductivity λ_{DC} was found by measuring the electrical conductivity λ_{AC} as a function of frequency f using an oscilloscopic technique (a version of impedance measurements when a known alternating current (AC) voltage signal V_{AC} of frequency f is applied across a sample, and the current through a sample I_{AC} is measured; in this case, V_{AC}, I_{AC}, and a phase lag between them are measured using an oscilloscope. This experimental setup allows us to perform the same type of electrical measurements as a standard impedance analyzer (Figure 1)). By knowing geometric parameters of the cell (its thickness and area), both dielectric permittivity and electrical conductivity can be found [111]. In the intermediate range of frequencies ($\sim 10^2$–10^4 Hz), the dependence of the electrical conductivity on frequency can be approximated by a power law known as Jonscher's power law or the universal dielectric response:

$$\lambda_{AC} = \lambda_{DC} + A f^m \qquad (1)$$

where A and m are empirical parameters [42,111].

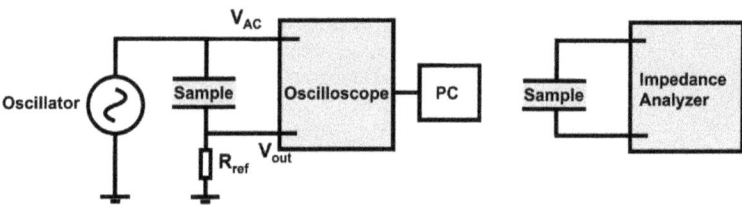

Figure 1. Schematic diagram of an experimental setup.

An analysis of the reported experimental results [38,52,110] was performed in the framework of the Langmuir adsorption model applied recently to molecular liquid crystals containing ions and nanoparticles [109]. An essential feature of this model is the consideration of the possibility of ionic contamination of the cell substrate and/or nanoparticles [109]. The details of this approach, its limitations, and its applicability to describe molecular liquid crystals with ions and nanomaterials can be found in recent papers [109,113].

3. Time-Dependent DC Electrical Conductivity of Liquid Crystal Cells

The first case study is related to the role of interactions between ions and substrates of the liquid crystal cell. In general, the liquid crystal substrates can act as a source of ions (if they are contaminated with ions) or they can capture ions already present in liquid crystals prior to filling an empty cell. In any real situation, we can expect a competition between ion-capturing and ion-releasing processes [113]. Experimental results reported in paper [110] allow us to analyze this competition between ionic processes in greater detail. According to [110], an empty cell was irradiated with a UV light prior to filling it with nematic liquid crystals 5CB. The source of UV light was a metal halide lamp (Panacol, Steinbach, Germany). The light intensity was 8 mW/cm^2. An ionizing UV light resulted in a generation of ions on the substrates of an empty liquid crystal cell. An empty cell exposed to a UV light was filled with nematic liquid crystals 5CB and the DC electrical conductivity was measured as a function of time (Figure 2a, circles). The observed non-monotonous time dependence of the DC electrical conductivity can be explained in terms of the competition between ion-releasing and ion-capturing processes [54]. Some ions already present in liquid crystals (prior to filling an empty cell) are captured by the substrates of a liquid crystal cell (Figure 2b, black curve (n_1)). At the same time, ions present on the substrates of a liquid crystal cell can be released into the liquid crystal bulk (Figure 2b, red curve (n_2)). The combination of ion-releasing and ion-capturing processes, shown in Figure 2b, results in an experimentally observed non-monotonous dependence $\lambda_{DC}(t)$ (Figure 2a).

Solid curves, shown in Figure 2, can be generated by applying Equations (2)–(4), briefly discussed below (additional details can be found in paper [113]).

The DC electrical conductivity λ_{DC} of molecular liquid crystals is defined by Equation (2):

$$\lambda_{DC} = \sum_i q_i \mu_i n_i \quad (2)$$

where $q_i = |e| = 1.6 \times 10^{-19} C$ ($i = 1, 2$), $n_i^+ = n_i^- = n_i$ is the volume concentration of ions of type i (we use index "1" ($i = 1$) to denote ions (fully dissociated ionic species) present in liquid crystals prior to filling an empty cell, and index "2" ($i = 2$) for ions generated by the substrates of a liquid crystal cell; $i = 1, 2$), and $\mu_i = \mu_i^+ + \mu_i^-$ is the effective ion mobility of ions of type i ($i = 1, 2$). Interactions between ions of type i and substrates are described by rate Equation (3):

$$\frac{dn_i}{dt} = -k_{Si}^{a\pm} n_i \frac{\sigma_{Si}}{d}\left(1 - \Theta_{S1}^{\pm} - \Theta_{S2}^{\pm}\right) + k_{Si}^{d\pm} \frac{\sigma_{Si}}{d} \Theta_{Si}^{\pm} \quad (3)$$

where the parameters $k_{Si}^{a\pm}$ and $k_{Si}^{d\pm}$ describe the time rate of ion-capturing and ion-releasing processes, respectively, and quantities $\Theta_{Si}^{\pm}$ stand for the fractional surface coverage of substrates by the i-th ions (i = 1, 2). Parameter σ_{Si} is the surface density of all surface sites on two substrates, and d is the cell thickness [113].

Equation (3) represents the conservation of the total number of ions of the i-th type:

$$n_{0i} + \frac{\sigma_{Si}}{d} \nu_{Si} = n_i + \frac{\sigma_{Si}}{d} \Theta_{Si}^{\pm} \quad (4)$$

where n_{0i} is the initial concentration of ions of the i-th type in liquid crystals, and ν_{Si} is the contamination factor of substrates [113].

Table 1. Parameters used to generate solid curves shown in Figure 2.

Physical Parameter	Value
n_{01}	9.36×10^{20} m^{-3}
n_{02}	0 m^{-3}
σ_{S1}	1.6×10^{18} m^{-2}
σ_{S2}	1.6×10^{18} m^{-2}
$k_{S1}^{d\pm}$	3.5×10^{-4} s^{-1}
$k_{S2}^{d\pm}$	1×10^{-3} s^{-1}
$K_{S1} = k_{S1}^{a\pm}/k_{S1}^{d\pm}$	8.5×10^{-24} m^3
$K_{S2} = k_{S2}^{a\pm}/k_{S2}^{d\pm}$	7.37×10^{-24} m^3
ν_{S1}	0
ν_{S2}	6.325×10^{-3}
μ_1	1.92×10^{-10} m^2/Vs
μ_2	2.1×10^{-10} m^2/Vs
d	13.5 μm

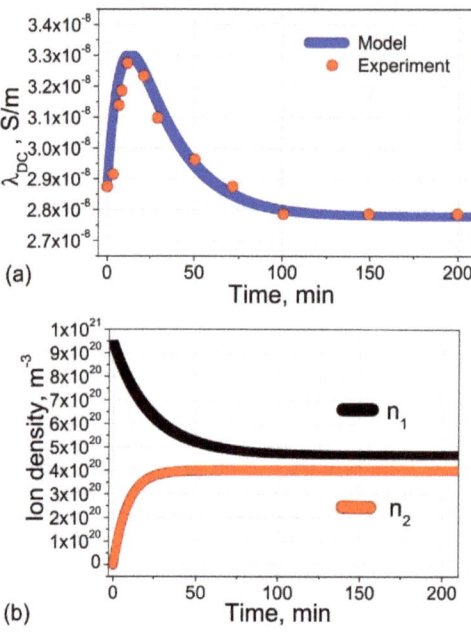

Figure 2. Time-dependent DC electrical conductivity and ion densities of nematic liquid crystals. The material parameters to generate solid curves are listed in Table 1. (**a**) Time-dependent DC electrical conductivity of a liquid crystal cell filled with nematic liquid crystals 5CB. The cell thickness is 13.5 µm. Red circles represent experimental datapoints reported in paper [110], and solid curve is a curve obtained by applying Equations (2)–(4). (**b**) The ion density of ions which were present in liquid crystals prior to filling an empty cell (n_1) and ions originated from the cell substrates (n_2) plotted as a function of time.

Solid curves, shown in Figure 2b, were generated by applying Equations (2)–(4) and using material parameters listed in Table 1. Figure 2b also reveals the presence of two characteristic time constants describing the ion-capturing ($\tau_1 = 23.7$ min) and ion-releasing ($\tau_2 = 8.9$ min) processes.

The results shown in Figure 2 suggest the adoption of the following experimental procedure for electrical measurements of liquid crystals. It is a good practice to record the time when an empty cell was filled with liquid crystals, and to measure the electrical conductivity as a function of time.

It should be noted that the model used to analyze the behavior of the DC electrical conductivity (Equations (2)–(4)) is elementary and has its limitations. Additional studies are required to uncover the mechanisms of ion generation in liquid crystal materials and liquid crystal substrates exposed to UV irradiation. Only a very limited number of publications can be found on this topic [114–116].

4. Steady-State DC Electrical Conductivity as a Function of the Cell Thickness

Equations (3) and (4) depend on the cell thickness. As a result, it can be expected that even a steady-state value of the DC electrical conductivity depends on the cell thickness. Indeed, the dependence of the DC electrical conductivity on the cell thickness was discussed in several papers [38,50]. Experimental results discussed in papers [38,50] were presented as a relative change in the DC electrical conductivity, $\Delta \lambda_{DC}/\lambda_{DC}$. Figure 3a shows an absolute value of the steady-state DC electrical conductivity of nematic liquid crystals MJ961190 plotted as a function of the cell thickness.

Table 2. Parameters used to generate solid curves shown in Figure 3.

Physical Parameter	Value
n_{01}	1.5×10^{19} m^{-3}
n_{02}	0 m^{-3}
σ_{S1}	3.6×10^{18} m^{-2}
σ_{S2}	0.75×10^{16} m^{-2}
$K_{S1} = k_{S1}^{a\pm}/k_{S1}^{d\pm}$	4.0×10^{-24} m^3
$K_{S2} = k_{S2}^{a\pm}/k_{S2}^{d\pm}$	5.0×10^{-24} m^3
ν_{S1}	0
ν_{S2}	2.5×10^{-2}
μ_1	2.5×10^{-9} m^2/Vs
μ_2	5.0×10^{-10} m^2/Vs

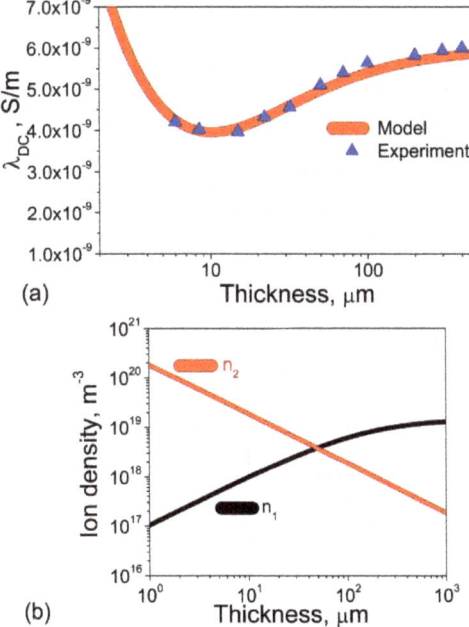

Figure 3. The dependence of the steady-state DC electrical conductivity and ion densities of nematic liquid crystals MJ961190 on the cell thickness. The material parameters to generate solid curves are listed in Table 2. (**a**) The dependence of the DC electrical conductivity of nematic liquid crystals MJ961190 on the cell thickness. (**b**) The dependence of the ion density of ions which were present in liquid crystals prior to filling an empty cell (n_1) and ions originated from the cell substrates (n_2) on the cell thickness.

The observed non-monotonous dependence $\lambda_{DC}(d)$ (Figure 3a) can be explained by considering the presence of two processes (ion capturing and ion releasing) modeled in Figure 3b. Ions present in liquid crystals prior to filling an empty cell (their ion density is n_1) are captured by the surface of the cell substrates. This ion-capturing effect becomes weaker (i.e., the concentration n_1 increases) as the cell gap increases (Figure 3b, black curve (n_1)). Once an empty cell is filled, ions originated from contaminated substrates are released into the bulk of liquid crystals. This ion-releasing effect is stronger in the case of thinner cells and becomes weaker if thicker cells are used (in other words, the ion density n_2 decreases when the cell gap increases (Figure 3b, red curve (n_2))).

Solid curves, shown in Figure 3, were generated by solving Equations (2)–(4), assuming steady-state conditions ($dn_i/dt = 0$). Material parameters are listed in Table 2.

It should be stressed that the range of available experimental datapoints shown in Figure 3 is limited. The smallest cell thickness is 4 μm. Even though theoretical curves show the dependence of the electrical conductivity starting from the thickness of 1 μm, an elementary model used in this paper can reach its limits and may require some modifications in the case of very thin cells. In addition, for thicker cells (>100 μm), the quality of alignment can affect the measured values of electrical conductivity because of its anisotropy. Even though all studied samples were inspected under crossed polarizers to verify their planar alignment, the alignment quality of thicker cells gradually decreases with increases in the cell thickness.

Figure 3 allows us to suggest the following improvements to existing experimental procedures for evaluations of the DC electrical conductivity. Even in a steady-state regime, the value of the DC electrical conductivity should be measured using cells of several thicknesses to obtain an idea about possible ionic processes in such systems.

5. Steady-State DC Electrical Conductivity of Nematic Liquid Crystals 6CB Doped with Iron Oxide Nanoparticles

If nanoparticles are used to modify the properties of liquid crystals, the evaluation of the DC electrical conductivity becomes more challenging because ions can interact with both nanoparticles and substrates of the cell. Moreover, nanoparticles can also act as a source of ions due to their uncontrolled ionic contamination [117]. Under certain conditions, nanomaterials contaminated with ions can behave in molecular liquid crystals in a similar way as weak electrolytes do [117]. As a result, to obtain deeper insight into the origin of ionic processes in such systems, it is important to measure the DC electrical conductivity as a function of the concentration of nanoparticles and as a function of the cell thickness. While the former aspect of experimental research was addressed in many publications (references to original publications can be found in topical review [31]), the latter issue remains practically unexplored [52]. Therefore, it is interesting and instructive to apply the same Langmuir adsorption model to analyze recently reported experimental data obtained for nematic liquid crystals 6CB doped with iron oxide nanoparticles [52].

In this case, to account for nanoparticle-induced ionic processes, Equations (3) and (4) can be modified by adding new terms responsible for both ion capturing and ion releasing by nanoparticles (Equations (5) and (6)) [113]:

$$\frac{dn_i}{dt} = -k_{Si}^{a\pm} n_i \frac{\sigma_{Si}}{d} \left(1 - \Theta_{S1}^{\pm} - \Theta_{S2}^{\pm}\right) + k_{Si}^{d\pm} \frac{\sigma_{Si}}{d} \Theta_{Si}^{\pm} - k_{NPi}^{a\pm} n_i n_{NP} A_{NP} \sigma_{Si}^{NP} \left(1 - \Theta_{NP1}^{\pm} - \Theta_{NP2}^{\pm}\right) + k_{NPi}^{d\pm} n_{NP} \sigma_{Si}^{NP} A_{NP} \Theta_{NPi}^{\pm} \quad (5)$$

$$n_{0i} + \frac{\sigma_{Si}}{d} \nu_{Si} + n_{NP} A_{NP} \sigma_{Si}^{NP} \nu_{NPi} = n_i + \frac{\sigma_{Si}}{d} \Theta_{Si}^{\pm} + n_{NP} A_{NP} \sigma_{Si}^{NP} \Theta_{NPi}^{\pm} \quad (6)$$

where the subscript "NP" refers to nanoparticles, and A_{NP} is the surface area of a single nanoparticle.

Experimental datapoints and theoretical curves computed by applying Equations (2), (5), and (6) are shown in Figure 4. Figure 4 indicates that for a given cell thickness, measurements of the DC electrical conductivity as a function of the concentration of nanoparticles can reveal whether nanomaterials act as ion-capturing or ion-generating objects. In the case shown in Figure 4, nanoparticles behave as ion-generating objects. It should be noted that the ion-releasing effect becomes more pronounced for thicker cells (Figure 4, red curve (50 μm thick cell)) and weakens if thinner cells are used (Figure 4, black curve (5 μm thick cell)). This fact can be explained in the following way. The substrates of a liquid crystal cell can capture only a limited number of ions released by nanoparticles. As a result, as the cell thickness increases, the ion density of ions brought by contaminated nanoparticles into a liquid crystal host also increases and leads to the increase in the electrical conductivity, as is evidenced from Figure 4.

Table 3. Parameters used to generate solid curves shown in Figures 4 and 5.

Physical Parameter	Value (5 μm Thick Cell)	Value (50 μm Thick Cell)
n_{01}	2.5×10^{20} m^{-3}	2.5×10^{20} m^{-3}
n_{02}	0 m^{-3}	0 m^{-3}
σ_{S1}	2.0×10^{18} m^{-2}	2.0×10^{18} m^{-2}
σ_{S2}	1.0×10^{18} m^{-2}	1.0×10^{18} m^{-2}
σ_{NP1}	2.0×10^{18} m^{-2}	2.0×10^{18} m^{-2}
σ_{NP2}	1.0×10^{18} m^{-2}	1.0×10^{18} m^{-2}
$K_{NP1} = k_{NP1}^{a\pm}/k_{NP1}^{d\pm}$	2.0×10^{-24} m^3	2.0×10^{-24} m^3
$K_{NP2} = k_{NP2}^{a\pm}/k_{NP2}^{d\pm}$	3.0×10^{-25} m^3	3.0×10^{-25} m^3
$K_{S1} = k_{S1}^{a\pm}/k_{S1}^{d\pm}$	1.0×10^{-24} m^3	1.0×10^{-24} m^3
$K_{S2} = k_{S2}^{a\pm}/k_{S2}^{d\pm}$	6.0×10^{-24} m^3	6.0×10^{-24} m^3
ν_{S1}	0	0
ν_{S2}	0	0
ν_{NP1}	0	0
ν_{NP2}	0.7×10^{-3}	3.2×10^{-3}
μ_1	6.32×10^{-10} m^2/Vs	6.32×10^{-10} m^2/Vs
μ_2	12.64×10^{-10} m^2/Vs	12.64×10^{-10} m^2/Vs
R_{NP}	2.5 nm	2.5 nm

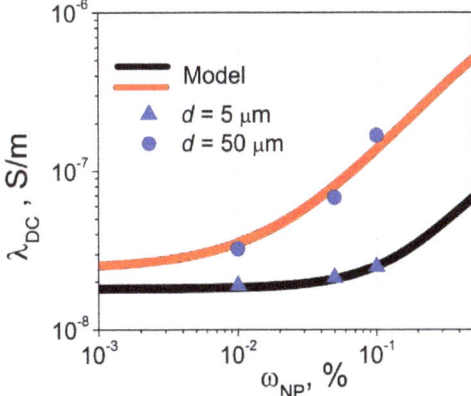

Figure 4. The dependence of the steady-state DC electrical conductivity of nematic liquid crystals 6CB on the weight concentration of iron oxide nanoparticles. The measurements were carried out using thin (5 μm) and thick (50 μm) cells. The material parameters to generate solid curves are listed in Table 3. Experimental datapoints are from paper [52].

Figure 5 offers additional insights into the ionic effects observed in liquid crystals 6CB doped with iron oxide nanoparticles. The concentration of ions which were present in liquid crystals prior to doping them with nanoparticles (n_1) decreases as the concentration of nanoparticles increases (Figure 5, black curves (n_1)). This is caused by the ion-capturing effect (nanoparticles capture ions). In addition, some fraction of such ions become trapped by the substrates of a cell. Figure 5a,b indicate that, in the case of ions which are inherently present in liquid crystals, the ion-capturing effect due to nanoparticles is much stronger than the ion-capturing effect due to the substrates for thicker cells (in this case, a noticeable decrease in the ion density n_1 begins at lower values of the weight concentration of nanoparticles ω_{NP}). Figure 5 also shows that nanoparticles also act as a source of a new type of ions in liquid crystals (Figure 5, red curves (n_2)). The concentration of ions generated by nanoparticles (n_2) increases as the concentration of nanodopants increases (Figure 5, red curves (n_2)). Interestingly, this increase becomes weaker if thinner cells are used because ions released by contaminated nanoparticles become trapped by the surfaces of the liquid crystal cell. According to Figure 5, in the case of a 50 µm-thick cell the concentration of ions generated by nanoparticles is nearly one order of magnitude greater than that in the case of a 5 µm-thick cell.

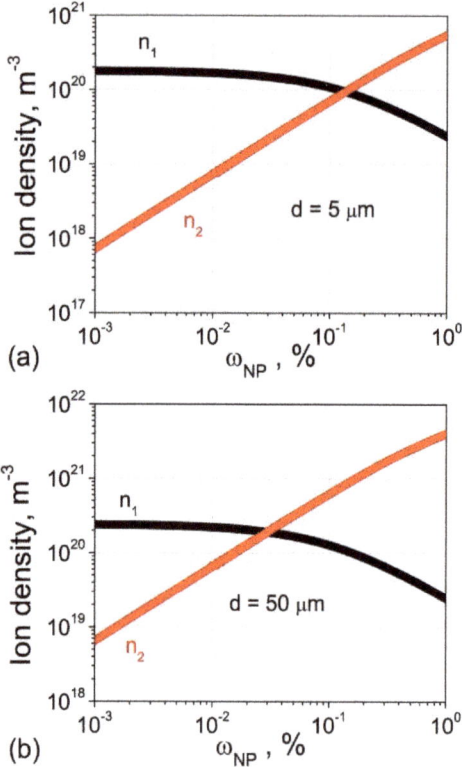

Figure 5. The dependence of the steady-state value of ion densities in nematic liquid crystals 6CB doped with iron oxide nanoparticles on the weight concentration of nanoparticles. The material parameters to generate solid curves are listed in Table 3. (**a**) The dependence of the ion density n_1 (ions which were present in liquid crystals prior to mixing them with nanoparticles) and of the ion density n_2 (ions originated in liquid crystals due to the ionic contamination of iron oxide nanoparticles) on the weight concentration of nanoparticles. The cell thickness is 5 µm (thin cell). (**b**) The dependence of the ion density n_1 (ions which were present in liquid crystals prior to mixing them with nanoparticles) and of the ion density n_2 (ions originated in liquid crystals due to the ionic contamination of iron oxide nanoparticles) on the weight concentration of nanoparticles. The cell thickness is 50 µm (thick cell).

It is interesting to comment on the values of material parameters used to generate solid curves shown in Figures 4 and 5. The values of all parameters listed in Table 3 are kept the same (for 5 µm and 50 µm thick cells) except for the value of the contamination factor of nanoparticles (its values it still on the order of 10^{-3} for both cases). Minor changes in the preparation process of liquid crystals doped with nanoparticles can result in some changes in the level of ionic contamination of nanoparticles.

Figures 4 and 5 can be used for improving experimental procedures for assessing DC electrical conductivity of liquid crystals doped with nanoparticles. The ion-capturing and ion-releasing processes caused by interactions between ions, substrates, and nanoparticles can result in dependence of the DC electrical conductivity on the cell thickness and on the concentration of nanoparticles. The use of relatively thin cells can hinder the strength of nanoparticle-induced ion-capturing or ion-releasing effect. To reveal the effect of nanoparticles on the electrical conductivity of liquid crystals not affected by the substrates, thicker cells should be used.

6. Conclusions

The results presented in this paper have important practical implications. If an experimentalist is aimed at improving experimental procedures for evaluating the DC electrical conductivity of liquid crystal materials several factors should be considered. In the case of undoped liquid crystal materials, interactions between ions and substrates of a liquid crystal cell can results in time-dependent electrical conductivity (Figure 2). Moreover, the DC electrical conductivity also depends on the cell thickness even if a steady state is reached (Figure 3). Therefore, it is advisable to record the time when an empty cell is filled with liquid crystal materials and measure the electrical conductivity as a function of time until a steady state is reached. Such measurements can provide valuable information about ionic processes in liquid crystal materials and to what degree the DC electrical conductivity is affected by the cell substrates. Time-dependent experiments can also reveal whether the cell substrates act as a source of ion generation or a source of ion capturing. Additional insight into the ionic phenomena in liquid crystals can be obtained by measuring the dependence of the DC electrical conductivity on the cell thickness. Such measurements can identify a range of cell thicknesses corresponding to the situation when interactions between ions and cell substrates can alter the DC electrical conductivity of liquid crystals and the range of cell thicknesses which do not alter the DC electrical conductivity.

In the case of liquid crystals doped with nanomaterials, a proper evaluation of the DC electrical conductivity of such systems can be obtained by varying both the concentration of nanoparticles and the cell thickness (Figures 4 and 5). To understand the variations in the measured values of the DC electrical conductivity, interactions between ions, nanoparticles, and substrates of liquid crystal cells should be considered. It is important to realize the dual (both ion capturing and ion releasing) role played by the substrates and nanoparticles. The use of thin cells can hinder the effect of nanoparticles on the electrical conductivity of liquid crystals. To mitigate this effect, thicker cells can be used. At the same time, the alignment quality should also be considered because it can set a limit on the maximum value of the cell thickness.

An adoption of the proposed suggestions could help reduce or even eliminate ambiguities in electrical measurements of liquid crystal materials performed by independent research groups around the globe.

Author Contributions: Conceptualization, O.V.K. and Y.G.; methodology, O.V.K., T.M.K. and Y.G.; software, Y.G.; validation, O.V.K., T.M.K. and Y.G.; formal analysis, O.V.K., T.M.K. and Y.G.; investigation, O.V.K., T.M.K. and Y.G.; resources, O.V.K. and Y.G.; data curation, Y.G.; writing—original draft preparation, Y.G.; writing—review and editing, O.V.K., T.M.K. and Y.G.; visualization, O.V.K., T.M.K. and Y.G.; supervision, O.V.K. and Y.G.; project administration, Y.G.; funding acquisition, Y.G. All authors have read and agreed to the published version of the manuscript.

Funding: This research was funded by the 2023–2024 CSU—AAUP Faculty Research Grant and the NASA CT Space Grant.

Data Availability Statement: All data that support the findings of this study are included within the article.

Acknowledgments: The authors would like to acknowledge funding from the 2023–2024 CSU—AAUP Faculty Research Grant and the NASA CT Space Grant.

Conflicts of Interest: The authors declare no conflict of interest.

References

1. Koide, N. (Ed.). 50 years of liquid crystal R&D that lead the way to the future. In *The Liquid Crystal Display Story*; Springer: Tokyo, Japan, 2014.
2. Jones, C. The fiftieth anniversary of the liquid crystal display. *Liq. Cryst. Today* **2018**, *27*, 44–70. [CrossRef]
3. Xiong, J.; Hsiang, E.-L.; He, Z.; Zhan, T.; Wu, S.-T. Augmented reality and virtual reality displays: Emerging technologies and future perspectives. *Light Sci. Appl.* **2021**, *10*, 216. [CrossRef] [PubMed]
4. Wang, Y.-J.; Lin, Y.-H. Liquid crystal technology for vergence-accommodation conflicts in augmented reality and virtual reality systems: A review. *Liq. Cryst. Rev.* **2021**, *9*, 35–64. [CrossRef]
5. Algorri, J.F.; Zografopoulos, D.C.; Urruchi, V.; Sánchez-Pena, J.M. Recent Advances in Adaptive Liquid Crystal Lenses. *Crystals* **2019**, *9*, 272. [CrossRef]
6. Lin, Y.; Wang, Y.; Reshetnyak, V. Liquid crystal lenses with tunable focal length. *Liq. Cryst. Rev.* **2017**, *5*, 111–143. [CrossRef]
7. d'Alessandro, A.; Asquini, R. Light Propagation in Confined Nematic Liquid Crystals and Device Applications. *Appl. Sci.* **2021**, *11*, 8713. [CrossRef]
8. Shin, Y.; Jiang, Y.; Wang, Q.; Zhou, Z.; Qin, G.; Yang, D.K. Flexoelectric-effect-based light waveguide liquid crystal display for transparent display. *Photon. Res.* **2022**, *10*, 407–414. [CrossRef]
9. Abdulhalim, I. Non-display bio-optic applications of liquid crystals. *Liq. Cryst. Today* **2011**, *20*, 44–60. [CrossRef]
10. Chigrinov, V.G. *Liquid Crystal Photonics*; Nova Science Pub Inc.: New York, NY, USA, 2014; 204p.
11. Otón, J.M.; Otón, E.; Quintana, X.; Geday, M.A. Liquid-crystal phase-only devices. *J. Mol. Liq.* **2018**, *267*, 469–483. [CrossRef]
12. De Sio, L.; Roberts, D.E.; Liao, Z.; Hwang, J.; Tabiryan, N.; Steeves, D.M.; Kimball, B.R. Beam shaping diffractive wave plates. *Appl. Opt.* **2018**, *57*, A118–A121. [CrossRef]
13. Morris, R.; Jones, C.; Nagaraj, M. Liquid Crystal Devices for Beam Steering Applications. *Micromachines* **2021**, *12*, 247. [CrossRef]
14. Geis, M.W.; Bos, P.J.; Liberman, V.; Rothschild, M. Broadband optical switch based on liquid crystal dynamic scattering. *Opt. Express* **2016**, *24*, 13812–13823. [CrossRef] [PubMed]
15. He, Z.; Zeng, J.; Zhu, S.; Zhang, D.; Ma, C.; Zhang, C.; Yu, P.; Miao, Z. A bistable light shutter based on polymer stabilized cholesteric liquid crystals. *Opt. Mater.* **2023**, *136*, 113426. [CrossRef]
16. Sung, G.-F.; Wu, P.-C.; Zyryanov, V.Y.; Lee, W. Electrically active and thermally passive liquid-crystal device toward smart glass. *Photon. Res.* **2021**, *9*, 2288–2295. [CrossRef]
17. Luo, L.; Liang, Y.; Feng, Y.; Mo, D.; Zhang, Y.; Chen, J. Recent Progress on Preparation Strategies of Liquid Crystal Smart Windows. *Crystals* **2022**, *12*, 1426. [CrossRef]
18. Liu, S.; Xu, F.; Zhan, J.; Qiang, J.; Xie, Q.; Yang, L.; Deng, S.; Zhang, Y. Terahertz liquid crystal programmable metasurface based on resonance switching. *Opt. Lett.* **2022**, *47*, 1891–1894. [CrossRef] [PubMed]
19. Chen, W.T.; Zhu, A.Y.; Capasso, F. Flat optics with dispersion-engineered metasurfaces. *Nat. Rev. Mater.* **2020**, *5*, 604–620. [CrossRef]
20. Jeng, S.C. Applications of Tamm plasmon-liquid crystal devices. *Liq. Cryst.* **2020**, *47*, 1223–1231. [CrossRef]
21. Chiang, W.; Silalahi, H.; Chiang, Y.C.; Hsu, M.C.; Zhang, Y.S.; Liu, J.H.; Yu, Y.; Lee, C.R.; Huang, C.Y. Continuously tunable intensity modulators with large switching contrasts using liquid crystal elastomer films that are deposited with terahertz metamaterials. *Opt. Express* **2020**, *28*, 27676–27687. [CrossRef]
22. Pozhidaev, E.P.; Minchenko, M.V.; Kuznetsov, A.V.; Tkachenko, T.P.; Barbashov, V.A. Broad temperature range ferrielectric liquid crystal as a highly sensitive quadratic electro-optical material. *Opt. Lett.* **2022**, *47*, 1598–1601. [CrossRef]
23. Lu, H.; Shi, J.; Wang, Q.; Xue, Y.; Yang, L.; Xu, M.; Zhu, J.; Qiu, L.; Ding, Y.; Zhang, J. Tunable multi-mode laser based on robust cholesteric liquid crystal microdroplet. *Opt. Lett.* **2021**, *46*, 5067–5070. [CrossRef]
24. Schenning, A.P.H.J.; Crawford, G.P.; Broer, D.J. (Eds.). *Liquid Crystal Sensors*; CRC Press, Taylor & Francis Group: Boca Raton, FL, USA, 2018; 164p.
25. Camley, R.; Celinski, Z.; Garbovskiy, Y.; Glushchenko, A. Liquid crystals for signal processing applications in the microwave and millimeter wave frequency ranges. *Liq. Cryst. Rev.* **2018**, *6*, 17–52. [CrossRef]
26. Jakoby, R.; Gaebler, A.; Weickhmann, C. Microwave Liquid Crystal Enabling Technology for Electronically Steerable Antennas in SATCOM and 5G Millimeter-Wave Systems. *Crystals* **2020**, *10*, 514. [CrossRef]
27. Ma, J.; Choi, J.; Park, S.; Kong, I.; Kim, D.; Lee, C.; Youn, Y.; Hwang, M.; Oh, S.; Hong, W.; et al. Liquid Crystals for Advanced Smart Devices with Microwave and Millimeter-Wave Applications: Recent Progress for Next-generation Communications. *Adv. Mater.* **2023**, 2302474, *accepted author manuscript*. [CrossRef]
28. Blinov, L.M. *Structure and Properties of Liquid Crystals*; Springer: New York, NY, USA, 2010.
29. Neyts, K.; Beunis, F. Ion Transport in Liquid Crystals. In *Handbook of Liquid Crystals: Physical Properties and Phase Behavior of Liquid Crystals*; Wiley-VCH: Weinheim, Germany, 2014; Volume 2, Chapter 11; pp. 357–382.

30. Éber, N.; Salamon, P.; Buka, Á. Electrically induced patterns in nematics and how to avoid them. *Liq. Cryst. Rev.* **2016**, *4*, 101–134. [CrossRef]
31. Garbovskiy, Y. Conventional and unconventional ionic phenomena in tunable soft materials made of liquid crystals and nanoparticles. *Nano Express* **2021**, *2*, 012004. [CrossRef]
32. Naemura, S. Electrical properties of liquid crystal materials for display applications. *Mater. Res. Soc. Symp. Proc.* **1999**, *559*, 263–274. [CrossRef]
33. Goodby, J.W.; Cowling, S.J. Conception, Discovery, Invention, Serendipity and Consortia: Cyanobiphenyls and Beyond. *Crystals* **2022**, *12*, 825. [CrossRef]
34. Kumari, P.; Basnet, B.; Wang, H.; Lavrentovich, O.D. Ferroelectric nematic liquids with conics. *Nat. Commun.* **2023**, *14*, 748. [CrossRef] [PubMed]
35. Lee, K.M.; Bunning, T.J.; White, T.J.; McConney, M.E.; Godman, N.P. Effect of Ion Concentration on the Electro-Optic Response in Polymer-Stabilized Cholesteric Liquid Crystals. *Crystals* **2021**, *11*, 7. [CrossRef]
36. Lee, K.M.; Marsh, Z.M.; Crenshaw, E.P.; Tohgha, U.N.; Ambulo, C.P.; Wolf, S.M.; Carothers, K.J.; Limburg, H.N.; McConney, M.E.; Godman, N.P. Recent Advances in Electro-Optic Response of Polymer-Stabilized Cholesteric Liquid Crystals. *Materials* **2023**, *16*, 2248. [CrossRef]
37. Colpaert, C.; Maximus, B.; Meyere, D. Adequate measuring techniques for ions in liquid crystal layers. *Liq. Cryst.* **1996**, *21*, 133–142. [CrossRef]
38. Kovalchuk, O.V.; Glushchenko, A.; Garbovskiy, Y. Improving experimental procedures for assessing electrical properties of advanced liquid crystal materials. *Liq. Cryst.* **2023**, *50*, 140. [CrossRef]
39. Barbero, G.; Evangelista, L.R. *Adsorption Phenomena and Anchoring Energy in Nematic Liquid Crystals*; Taylor & Francis: Boca Raton, FL, USA, 2006.
40. Khazimullin, M.V.; Lebedev, Y.A. Influence of dielectric layers on estimates of diffusion coefficients and concentrations of ions from impedance spectroscopy. *Phys. Rev. E* **2019**, *100*, 062601. [CrossRef] [PubMed]
41. Karaawi, A.R.; Gavrilyak, M.V.; Boronin, V.A.; Gavrilyak, A.M.; Kazachonok, J.V.; Podgornov, F.V. Direct current electric conductivity of ferroelectric liquid crystals–gold nanoparticles dispersion measured with capacitive current technique. *Liq. Cryst.* **2020**, *47*, 1507–1515. [CrossRef]
42. Barrera, A.; Binet, C.; Dubois, F.; Hébert, P.-A.; Supiot, P.; Foissac, C.; Maschke, U. Temperature and frequency dependence on dielectric permittivity and electrical conductivity of recycled Liquid Crystals. *J. Mol. Liq.* **2023**, *378*, 121572. [CrossRef]
43. Vaxiviere, J.; Labroo, B.; Martinot-Lagarde, P. Ion Bump in the Ferroelectric Liquid Crystal Domains Reversal Current. *Mol. Cryst. Liq. Cryst. Inc. Nonlinear Opt.* **1989**, *173*, 61–73. [CrossRef]
44. Sugimura, A.; Matsui, N.; Takahashi, Y.; Sonomura, H.; Naito, H.; Okuda, M. Transient currents in nematic liquid crystals. *Phys. Rev. B* **1991**, *43*, 8272–8276. [CrossRef] [PubMed]
45. Inoue, M. Review of various measurement methodologies of migration ion influence on LCD image quality and new measurement proposal beyond LCD materials. *J. Soc. Inf. Disp.* **2020**, *28*, 92–110. [CrossRef]
46. Mizusaki, M.; Ishihara, S. A Novel Technique for Determination of Residual Direct-Current Voltage of Liquid Crystal Cells with Vertical and In-Plane Electric Fields. *Symmetry* **2021**, *13*, 816. [CrossRef]
47. Sasaki, N. A new measurement method for ion density in TFT-LCD panels. *Mol. Cryst. Liq. Cryst. Sci. Technol. Sect. A Mol. Cryst. Liq. Cryst.* **2001**, *367*, 671–679. [CrossRef]
48. Dhara, S.; Madhusudana, N.V. Ionic contribution to the dielectric properties of a nematic liquid crystal in thin cells. *J. Appl. Phys.* **2001**, *90*, 3483–3488. [CrossRef]
49. Kumar, A.; Varshney, D.; Prakash, J. Role of ionic contribution in dielectric behaviour of a nematic liquid crystal with variable cell thickness. *J. Mol. Liq.* **2020**, *303*, 112520. [CrossRef]
50. Kovalchuk, O.V. Adsorption of ions and thickness dependence of conductivity in liquid crystal. *Semicond. Phys. Quantum Electron. Optoelectron.* **2011**, *14*, 452–455. [CrossRef]
51. Shukla, R.K.; Chaudhary, A.; Bubnov, A.; Raina, K.K. Multiwalled carbon nanotubes-ferroelectric liquid crystal nanocomposites: Effect of cell thickness and dopant concentration on electro-optic and dielectric behaviour. *Liq. Cryst.* **2018**, *45*, 1672–1681. [CrossRef]
52. Kovalchuk, O.; Kovalchuk, T.; Tomašovičová, N.; Timko, M.; Zakutanska, K.; Miakota, D.; Kopčanský, P.; Shevchuk, O.; Garbovskiy, Y. Dielectric and electrical properties of nematic liquid crystals 6CB doped with iron oxide nanoparticles. The combined effect of nanodopant concentration and cell thickness. *J. Mol. Liq.* **2022**, *366*, 120305. [CrossRef]
53. Garbovskiy, Y. Ions and size effects in nanoparticle/liquid crystal colloids sandwiched between two substrates. The Case of Two Types Fully Ionized Species. *Chem. Phys. Lett.* **2017**, *679*, 77–85. [CrossRef]
54. Garbovskiy, Y. Time-dependent electrical properties of liquid crystal cells: Unravelling the origin of ion generation. *Liq. Cryst.* **2018**, *45*, 1540–1548. [CrossRef]
55. Webb, D.; Garbovskiy, Y. Overlooked Ionic Phenomena Affecting the Electrical Conductivity of Liquid Crystals. *Eng. Proc.* **2021**, *11*, 1. [CrossRef]
56. Webb, D.; Garbovskiy, Y. Steady-State and Transient Electrical Properties of Liquid Crystal Cells. *Chem. Proc.* **2022**, *9*, 15. [CrossRef]

57. Lagerwall, J.P.F.; Scalia, G. (Eds.). *Liquid Crystals with Nano and Microparticles*; World Scientific: Singapore, 2016; Volume 2; ISBN 978-981-4619-25-7.
58. Shen, Y.; Dierking, I. Perspectives in Liquid-Crystal-Aided Nanotechnology and Nanoscience. *Appl. Sci.* **2019**, *9*, 2512. [CrossRef]
59. Dierking, I. Nanomaterials in Liquid Crystals. *Nanomaterials* **2018**, *8*, 453. [CrossRef] [PubMed]
60. Lee, W.; Kumar, S. *Unconventional Liquid Crystals and Their Applications*; De Gruyter: Berlin, Germany; Boston, MA, USA, 2021.
61. Rastogi, A.; Mishra, A.; Pandey, F.P.; Manohar, R.; Parmar, A.S. Enhancing physical characteristics of thermotropic nematic liquid crystals by dispersing in various nanoparticles and their potential applications. *Emergent Mater.* **2023**, *6*, 101. [CrossRef]
62. Shukla, R.K.; Liebig, C.M.; Evans, D.R.; Haase, W. Electro-optical behaviour and dielectric dynamics of harvested ferroelectric $LiNbO_3$ nanoparticle-doped ferroelectric liquid crystal nanocolloids. *RSC Adv.* **2014**, *4*, 18529–18536. [CrossRef]
63. Basu, R.; Garvey, A. Effects of ferroelectric nanoparticles on ion transport in a liquid crystal. *Appl. Phys. Lett.* **2014**, *105*, 151905. [CrossRef]
64. Hsiao, Y.G.; Huang, S.M.; Yeh, E.R.; Lee, W. Temperature-dependent electrical and dielectric properties of nematic liquid crystals doped with ferroelectric particles. *Displays* **2016**, *44*, 61–65. [CrossRef]
65. Al-Zangana, S.; Turner, M.; Dierking, I. A comparison between size dependent paraelectric and ferroelectric $BaTiO_3$ nanoparticle doped nematic and ferroelectric liquid crystals. *J. Appl. Phys.* **2017**, *121*, 085105. [CrossRef]
66. Kumar, P.; Debnath, S.; Rao, N.V.S.; Sinha, A. Nanodoping: A route for enhancing electro-optic performance of bent core nematic system. *J. Phys. Condens. Matter.* **2018**, *30*, 095101. [CrossRef]
67. Shoarinejad, S.; Mohammadi Siahboomi, R. Ordering behavior and electric response of a ferroelectric nano-doped liquid crystal with ion impurity effects. *J. Appl. Phys.* **2021**, *129*, 025101. [CrossRef]
68. Lalik, S.; Deptuch, A.; Jaworska-Gołąb, T.; Fryń, P.; Dardas, D.; Stefańczyk, O.; Urbańska, M.; Marzec, M. Modification of AFLC Physical Properties by Doping with $BaTiO_3$ Particles. *J. Phys. Chem. B* **2020**, *124*, 6055. [CrossRef] [PubMed]
69. Łoś, J.; Drozd-Rzoska, A.; Rzoska, S.J. Critical-like behavior of low-frequency dielectric properties in compressed liquid crystalline octyloxycyanobiphenyl (8OCB) and its nanocolloid with paraelectric $BaTiO_3$. *J. Mol. Liq.* **2023**, *377*, 121555. [CrossRef]
70. Salah, M.B.; Nasri, R.; Alharbi, A.N.; Althagafi, T.M.; Soltani, T. Thermotropic liquid crystal doped with ferroelectric nanoparticles: Electrical behavior and ion trapping phenomenon. *J. Mol. Liq.* **2022**, *357*, 119142. [CrossRef]
71. Mertelj, A.; Lisjak, D. Ferromagnetic nematic liquid crystals. *Liq. Cryst. Rev.* **2017**, *5*, 1–33. [CrossRef]
72. Studenyak, I.P.; Kopčanský, P.; Timko, M.; Mitroova, Z.; Kovalchuk, O.V. Effects of non-additive conductivity variation for a nematic liquid crystal caused by magnetite and carbon nanotubes at various scales. *Liq. Cryst.* **2017**, *44*, 1709. [CrossRef]
73. Gao, L.; Dai, Y.; Li, T.; Tang, Z.; Zhao, X.; Li, Z.; Meng, X.; He, Z.; Li, J.; Cai, M.; et al. Enhancement of Image Quality in LCD by Doping γ-Fe_2O_3 Nanoparticles and Reducing Friction Torque Difference. *Nanomaterials* **2018**, *8*, 911. [CrossRef]
74. Jessy, P.J.; Radha, R.; Patel, N. Highly improved dielectric behaviour of ferronematic nanocomposite for display application. *Liq. Cryst.* **2019**, *46*, 772.
75. Dalir, N.; Javadian, S. Evolution of morphology and electrochemical properties of colloidal nematic liquid crystal doped with carbon nanotubes and magnetite. *J. Mol. Liq.* **2019**, *287*, 110927. [CrossRef]
76. Meng, X.; Li, J.; Lin, Y.; Liu, X.; Liu, N.; Ye, W.; Li, D.; He, Z. Polymer dispersed liquid crystals doped with low concentration γ-Fe_2O_3 nanoparticles. *Liq. Cryst.* **2021**, *48*, 1791. [CrossRef]
77. Meng, X.; Li, J.; Lin, Y.; Liu, X.; Li, G.; Zhao, J.; Miao, Y.; Li, W.; Ye, W.; Li, D.; et al. Electro-optical response of polymer-dispersed liquid crystals doped with γ-Fe_2O_3 nanoparticles. *Liq. Cryst.* **2022**, *49*, 855. [CrossRef]
78. Meng, X.; Li, J.; Lin, Y.; Liu, X.; Zhao, J.; Li, D.; He, Z. Optimization approach for the dilute magnetic polymer-dispersed liquid crystal. *Opt. Mater.* **2022**, *131*, 112670. [CrossRef]
79. Shcherbinin, D.; Konshina, E. Ionic impurities in nematic liquid crystal doped with quantum dots CdSe/ZnS. *Liq. Cryst.* **2017**, *44*, 648. [CrossRef]
80. Shcherbinin, D.P.; Konshina, E.A. Impact of titanium dioxide nanoparticles on purification and contamination of nematic liquid crystals. *Beilstein J. Nanotechnol.* **2017**, *8*, 2766–2770. [CrossRef]
81. Yadav, G.; Katiyar, R.; Pathak, G.; Manohar, R. Effect of ion trapping behavior of TiO_2 nanoparticles on different parameters of weakly polar nematic liquid crystal. *J. Theor. Appl. Phys.* **2018**, *12*, 191. [CrossRef]
82. Konshina, E.; Shcherbinin, D.; Kurochkina, M. Comparison of the properties of nematic liquid crystals doped with TiO_2 and CdSe/ZnS nanoparticles. *J. Mol. Liq.* **2018**, *267*, 308–314. [CrossRef]
83. Rastogi, A.; Agrahari, K.; Pathak, G.; Srivastava, A.; Herman, J.; Manohar, R. Study of an interesting physical mechanism of memory effect in nematic liquid crystal dispersed with quantum dots. *Liq. Cryst.* **2019**, *46*, 725. [CrossRef]
84. Prakash, J.; Khan, S.; Chauhan, S.; Biradar, A. Metal oxide-nanoparticles and liquid crystal composites: A review of recent progress. *J. Mol. Liq.* **2020**, *297*, 112052. [CrossRef]
85. Seidalilir, Z.; Soheyli, E.; Sabaeian, M.; Sahraei, R. Enhanced electrochemical and electro-optical properties of nematic liquid crystal doped with Ni:ZnCdS/ZnS core/shell quantum dots. *J. Mol. Liq.* **2020**, *320*, 114373. [CrossRef]
86. Rani, A.; Chakraborty, S.; Sinha, A. Effect of CdSe/ZnS quantum dots doping on the ion transport behavior in nematic liquid crystal. *J. Mol. Liq.* **2021**, *342*, 117327. [CrossRef]
87. Chauhan, G.; Malik, P.; Deep, A. Morphological, dielectric, electro-optic and photoluminescence properties of titanium oxide nanoparticles enriched polymer stabilized cholesteric liquid crystal composites. *J. Mol. Liq.* **2023**, *376*, 121406. [CrossRef]

88. Lisetski, L.; Minenko, S.; Samoilov, A.; Lebovka, N. Optical density and microstructure-related properties of photoactive nematic and cholesteric liquid crystal colloids with carbon nanotubes. *J. Mol. Liq.* **2017**, *235*, 90. [CrossRef]
89. Tomylko, S.; Yaroshchuk, O.; Koval'chuk, O.; Lebovka, N. Structural evolution and dielectric properties of suspensions of carbon nanotubes in nematic liquid crystals. *Phys. Chem. Chem. Phys.* **2017**, *19*, 16456. [CrossRef] [PubMed]
90. Basu, R.; Lee, A. Ion trapping by the graphene electrode in a graphene-ITO hybrid liquid crystal cell. *Appl. Phys. Lett.* **2017**, *111*, 161905. [CrossRef]
91. Cetinkaya, M.; Yildiz, S.; Ozbek, H. The effect of -COOH functionalized carbon nanotube doping on electro-optical, thermo-optical and elastic properties of a highly polar smectic liquid crystal. *J. Mol. Liq.* **2018**, *272*, 801. [CrossRef]
92. Shukla, R.K.; Chaudhary, A.; Bubnov, A.; Hamplova, V.; Raina, K.K. Electrically switchable birefringent self-assembled nanocomposites: Ferroelectric liquid crystal doped with the multiwall carbon nanotubes. *Liq. Cryst.* **2020**, *47*, 1379. [CrossRef]
93. Barrera, A.; Binet, C.; Dubois, F.; Hébert, P.-A.; Supiot, P.; Foissac, C.; Maschke, U. Dielectric Spectroscopy Analysis of Liquid Crystals Recovered from End-of-Life Liquid Crystal Displays. *Molecules* **2021**, *26*, 2873. [CrossRef] [PubMed]
94. Barrera, A.; Binet, C.; Dubois, F.; Hébert, P.-A.; Supiot, P.; Foissac, C.; Maschke, U. Recycling of liquid crystals from e-waste. *Detritus* **2022**, *55*, 55–61. [CrossRef]
95. Kumar Singh, P.; Dubey, P.; Dhar, R.; Dabrowski, R. Functionalized and non-functionalized multi walled carbon nanotubes in the anisotropic media of liquid crystalline material. *J. Mol. Liq.* **2023**, *369*, 120889. [CrossRef]
96. Chausov, D.; Kurilov, A.; Smirnova, A.; Stolbov, D.; Kucherov, R.; Emelyanenko, A.; Savilov, S.; Usol'tseva, N. Mesomorphism, dielectric permittivity, and ionic conductivity of cholesterol tridecylate doped with few-layer graphite fragments. *J. Mol. Liq.* **2023**, *374*, 121139. [CrossRef]
97. Shivaraja, S.; Sahai, M.; Gupta, R.; Manjuladevi, V. Superior electro-optical switching properties in polymer dispersed liquid crystals prepared with functionalized carbon nanotube nanocomposites of LC for switchable window applications. *Opt. Mater.* **2023**, *137*, 113546. [CrossRef]
98. Lisetski, L.; Bulavin, L.; Lebovka, N. Effects of Dispersed Carbon Nanotubes and Emerging Supramolecular Structures on Phase Transitions in Liquid Crystals: Physico-Chemical Aspects. *Liquids* **2023**, *3*, 246–277. [CrossRef]
99. Middha, M.; Kumar, R.; Raina, K. Photoluminescence tuning and electro-optical memory in chiral nematic liquid crystals doped with silver nanoparticles. *Liq. Cryst.* **2016**, *43*, 1002. [CrossRef]
100. Podgornov, F.V.; Wipf, R.; Stühn, B.; Ryzhkova, A.V.; Haase, W. Low-frequency relaxation modes in ferroelectric liquid crystal/gold nanoparticle dispersion: Impact of nanoparticle shape. *Liq. Cryst.* **2016**, *43*, 1536. [CrossRef]
101. Urbanski, M.; Lagerwall, J.P.F. Why organically functionalized nanoparticles increase the electrical conductivity of nematic liquid crystal dispersions. *J. Mater. Chem. C* **2017**, *5*, 8802. [CrossRef]
102. Yan, X.; Zhou, Y.; Liu, W.; Liu, S.; Hu, X.; Zhao, W.; Zhou, G.; Yuan, D. Effects of silver nanoparticle doping on the electro-optical properties of polymer stabilized liquid crystal devices. *Liq. Cryst.* **2020**, *47*, 1131. [CrossRef]
103. Shivaraja, S.J.; Gupta, R.K.; Kumar, S.; Manjuladevi, V. Enhanced electro-optical response of nematic liquid crystal doped with functionalized silver nanoparticles in twisted nematic configuration. *Liq. Cryst.* **2020**, *47*, 1678. [CrossRef]
104. Chausov, D.N.; Kurilov, A.D.; Kucherov, R.N.; Simakin, A.V.; Gudkov, S.V. Electro-optical performance of nematic liquid crystals doped with gold nanoparticles. *J. Phys. Condens. Matter* **2020**, *32*, 395102. [CrossRef]
105. Debnath, A.; Mandal, P.K.; Sarma, A.; Gutowski, O. Effect of silver nanoparticle doping on the physicochemical properties of a room temperature ferroelectric liquid crystal mixture. *J. Mol. Liq.* **2020**, *319*, 114185. [CrossRef]
106. Basu, R.; Gess, D.T. Ion trapping, reduced rotational viscosity, and accelerated electro-optic response characteristics in gold nano-urchin–nematic suspensions. *Phys. Rev. E* **2023**, *107*, 024705. [CrossRef]
107. Garbovskiy, Y. Nanomaterials in Liquid Crystals as Ion-Generating and Ion-Capturing Objects. *Crystals* **2018**, *8*, 264. [CrossRef]
108. Garbovskiy, Y. Nanoparticle—Enabled Ion Trapping and Ion Generation in Liquid Crystals. *Adv. Condens. Matter Phys.* **2018**, *2018*, 8914891. [CrossRef]
109. Garbovskiy, Y. A perspective on the Langmuir adsorption model applied to molecular liquid crystals containing ions and nanoparticles. *Front. Soft Matter.* **2022**, *2*, 1079063. [CrossRef]
110. Kravchuk, R.; Koval'chuk, O.; Yaroshchuk, O. Filling initiated processes in liquid crystal cell. *Mol. Cryst. Liq. Cryst.* **2002**, *384*, 111–119. [CrossRef]
111. Barsoukov, E.; Macdonald, J.R. *Impedance Spectroscopy: Theory, Experiment, and Applications*, 2nd ed.; Wiley-Interscience: Hoboken, NJ, USA, 2005.
112. Yaroshchuk, O.V.; Kiselev, A.D.; Kravchuk, R.M. Liquid-crystal anchoring transitions on aligning substrates processed by a plasma beam. *Phys. Rev. E* **2008**, *77*, 031706. [CrossRef] [PubMed]
113. Garbovskiy, Y. Kinetics of Ion-Capturing/Ion-Releasing Processes in Liquid Crystal Devices Utilizing Contaminated Nanoparticles and Alignment Films. *Nanomaterials* **2018**, *8*, 59. [CrossRef] [PubMed]
114. Lackner, A.M.; Margerum, J.D.; Van Ast, C. Near ultraviolet photostability of liquid crystal mixtures. *Mol. Cryst. Liq. Cryst.* **1986**, *141*, 289–310. [CrossRef]
115. Xu, H.; Davey, A.B.; Wilkinson, T.D.; Crossland, W.A.; Chapman, J.; Duffy, W.L.; Kelly, S.M. Performance of UV-stable STN Mixtures for PL-LCDs. *Mol. Cryst. Liq. Cryst.* **2004**, *411*, 79–91. [CrossRef]

116. Konovalov, V.; Fauchille, J.; Yakovenko, S. A lifetime model for LCOS panel under intense illumination. *J. SID* **2006**, *14*, 247–256. [CrossRef]
117. Garbovskiy, Y. On the Analogy between Electrolytes and Ion-Generating Nanomaterials in Liquid Crystals. *Nanomaterials* **2020**, *10*, 403. [CrossRef]

Disclaimer/Publisher's Note: The statements, opinions and data contained in all publications are solely those of the individual author(s) and contributor(s) and not of MDPI and/or the editor(s). MDPI and/or the editor(s) disclaim responsibility for any injury to people or property resulting from any ideas, methods, instructions or products referred to in the content.

Article

Phase-Only Liquid-Crystal-on-Silicon Spatial-Light-Modulator Uniformity Measurement with Improved Classical Polarimetric Method

Xinyue Zhang and Kun Li *

School of Electronic Science and Engineering, Southeast University, Nanjing 210096, China
* Correspondence: kl330@seu.edu.cn

Abstract: The classical polarimetric method has been widely used in liquid crystal on silicon (LCoS) phase measurement with a simple optical setup. However, due to interference caused by LCoS cover glass reflections, the method lacks accuracy for phase uniformity measurements. This paper is aimed at mathematically analyzing the errors caused by non-ideal glass reflections and proposing procedures to reduce or eliminate such errors. The measurement is discussed in three conditions, including the ideal condition with no reflections from the LCoS cover glass, the condition with only the front reflection from the cover glass, and the condition with only the back reflection from the cover glass. It is discovered that the backward reflection makes the largest contribution to the overall measurement error, and it is the main obstacle to high-quality measurements. Several procedures, including optical alignment, LC layer thickness measurement, and phase estimation method, are proposed, making the uniformity measurement more qualitative and consistent.

Keywords: liquid crystal on silicon; phase uniformity measurement; classical polarimetric method; interference

1. Introduction

Liquid crystal on silicon spatial light modulators (LCoS-SLMs) are devices used to perform spatial modulation of the light wavefront, and they have been commonly used in a wide range of applications, including holographic displays and others [1]. Phase-only LCoS-SLM requires an accurate spatial phase response, which describes the phase uniformities and linearity across the whole active area [1–3].

Multiple research groups [4–6] have presented interferometry methods to measure and calibrate the reflected wavefront from LCoS-SLMs, and such methods have become the preferred choice for measuring the LCoS phase uniformity whenever available. Figure 1 shows one of the possible optical setups for the interferometry measurement of the LCoS wavefront [5]. However, due to the nature of interferometry, these methods usually require a highly stable environment with minimal vibration and ambient light [7–9], which is not always available in every circumstance.

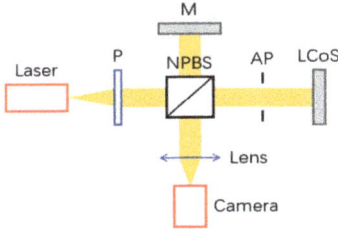

Figure 1. Interferometry method setup for phase uniformity measurement (P, polarizer, NPBS, non-polarizing beam splitter, AP, aperture, M, reference flat mirror).

In this paper, we evaluated the classical polarimetric method [10–13] to validate its application in the measurement of the non-uniform phase response of LCoS-SLMs. The method can measure the absolute phase retardation and phase flicker at any region of the active area, while it only requires basic equipment, as demonstrated in Figure 2, as well as a normal lab test environment without ambient light controlling. The classical polarimetric method is an indirect measurement of phase retardation. It measures the relation between output light intensity I and input gray level GL, and it calculates phase retardation Γ based on normalized intensity I_{norm}. Such a setup is not sensitive to vibrations, and by measuring the normalized light intensity in the output, the effect of ambient light is mostly eliminated.

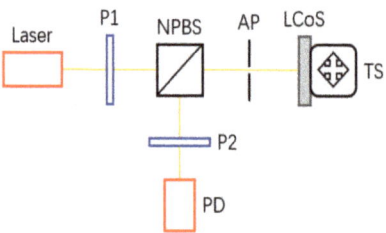

Figure 2. Basic optical setup for polarimetric method (P, polarizer, NPBS, non-polarizing beam splitter, AP, aperture, TS, translation stage, PD, photo detector).

The measuring process divides the whole active area of the tested LCoS device into a grid of regions, and in each region, the classical polarimetric method is performed to acquire its phase response [13]. A two-axis linear translation stage is used to traverse all regions. By combining the results from all regions, a distribution of the phase response could be calculated. Additionally, to ensure consistent measurements, the LCoS device was also kept at a stable temperature during the whole process.

However, there are some disagreements between ideal conditions in theory and actual test environments in practice, especially when anti-reflection (AR) coating is not present. Figure 3a shows an actual uniformity measurement from one of our LCoS devices with the setup shown in Figure 2, and the rippled result is clearly not what could be expected in an LCoS assembly [14,15], and it does not provide a solid foundation for the calibration process.

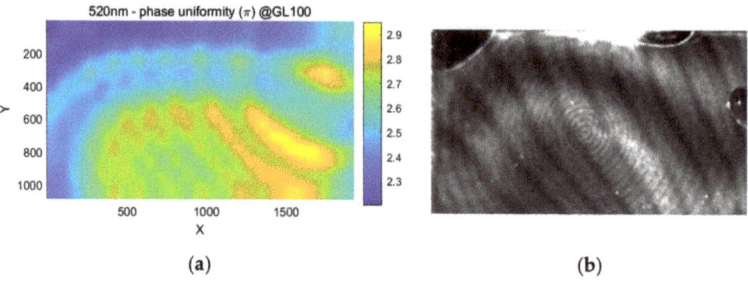

Figure 3. Problematic phase measurement due to non-ideal factors: (**a**) phase measurement result; (**b**) interference pattern of extraordinary light on the LCoS device.

During this experiment, we found that the pattern appearing in the phase result shared close similarity with the interference pattern observed on the LCoS device, shown in Figure 3b. According to this outcome, we were investigating a relationship between the error pattern and the reflections on the cover glass of the LCoS device with analytical modeling and simulation and comparing them with more detailed experiments.

Our work is aimed at analyzing and addressing the error caused by glass reflections to make the polarimetric method a more viable means for LCoS phase uniformity measurement. By considering the glass reflections and choosing unaffected data points, our improved measuring method could better compensate for the error caused by glass reflections, thus achieving a more accurate phase uniformity measurement compared to the original classical polarimetric method.

2. Materials and Methods

The polarimetric method used in this paper is based on the work by Márquez, A. et.al. [10,11], which used a single collimated laser beam passing through polarizer P1, LCoS (as a waveplate), and polarizer P2 in order, and it calculated the phase retardation by measuring the transmitted laser intensity. The basic principle of the polarimetric method is demonstrated in Figure 4, with the slow axis of the waveplate defined as the X-axis.

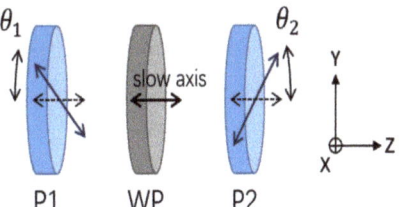

Figure 4. Principle of the polarimetric method (P, polarizer; WP, waveplate).

2.1. Basic Polarimetric Method

To obtain the relationship between the phase retardation of the LCoS device and the transmitted light intensity, some theoretical development is needed. The detailed derivation process with Jones calculus under ideal condition can be found at [10,11]. This sub-section only discusses the parameter definitions and the final conclusion as a foundation for later analysis.

First, the induced phase difference of ordinary light and extraordinary light after passing through the LCoS device are marked as φ_o and φ_{eff}, which correspond to their optical path lengths (OPLs) respectively. The liquid crystal (LC) is assumed to be a positive birefringence material, where $\varphi_{eff} > \varphi_o$ is always true.

The formulas below define φ_o, φ_{eff}, and phase retardation Γ, where $n_o(\lambda)$ and $n_{eff}(\lambda)$ are the effective refractive indices of LC material at the wavelength λ, and d is the distance that light travels through the liquid crystal, which is double the thickness of the LC layer in this LCoS device. Tilting of the LC molecule is considered in $n_{eff}(\lambda)$, which indicates the effective refractive index on the slow axis of the LCoS device.

$$\varphi_o = \frac{2\pi \cdot n_o(\lambda) \cdot d}{\lambda} \quad (1)$$

$$\varphi_{eff} = \frac{2\pi \cdot n_{eff}(\lambda) \cdot d}{\lambda} \quad (2)$$

$$\Gamma = \varphi_{eff} - \varphi_o = \frac{2\pi \cdot \left(n_{eff}(\lambda) - n_o(\lambda)\right) \cdot d}{\lambda} \quad (3)$$

For the simplicity of the calculation, the orientation of the slow axis of the LCoS device, which is the polarization orientation of the extraordinary light, is defined as the X axis. The Jones matrix of the LCoS device $W\left(\varphi_{eff}, \varphi_o\right)$ can be written as follows:

$$W\left(\varphi_{eff}, \varphi_o\right) = \begin{bmatrix} e^{i\varphi_{eff}} & 0 \\ 0 & e^{i\varphi_o} \end{bmatrix} \quad (4)$$

At the same time, the Jones matrix of the polarizer $P(\theta)$ can be written as follows, where θ is the angle between its transmissive axis and X-axis defined above.

$$P(\theta) = \begin{bmatrix} \cos^2\theta & \cos\theta\sin\theta \\ \cos\theta\sin\theta & \sin^2\theta \end{bmatrix}, \quad \theta \in \left(-\frac{\pi}{2}, \frac{\pi}{2}\right) \quad (5)$$

For the simplicity of the calculation again, the light intensity of the laser beam after polarizer P1 is marked as I_{in}. When θ_1 and θ_2 are orthogonal to each other, which means $\theta_1 - \theta_2 = \pm\frac{\pi}{2}$, I_{out} can be calculated as:

$$I_{out} = \frac{I_{in}}{2}\sin^2(2\theta_1) \cdot (1 - \cos\Gamma) \quad (6)$$

As can be seen, I_{out} can always be normalized into I_{norm}, which is only related to phase retardation Γ. As a result, Γ can be calculated from normalized light intensity I_{norm} during measurement:

$$I_{norm} = \frac{1}{2}(1 - \cos\Gamma) \quad (7)$$

$$\Gamma = \operatorname{acos}(1 - 2I_{norm}) \quad (8)$$

Under ideal conditions in which the reflection from the LCoS cover glass is ignored, and all optical components, including the laser beam and polarizers, are perfectly aligned, the measurement is guaranteed to be error-free.

2.2. Reflections on the LCoS Cover Glass

All the discussion and math formulas in Section 2.1 are based on an ideal modeling of the optical system. It is assumed that all the incident light is modulated by the LC layer exactly twice, including once inward and once outward. However, this case usually does not apply in real-life practice.

In practice, part of the incident light would be reflected without modulation from the LC layer, while part of the remaining light can be modulated multiple times [16]. Such un-modulated or over-modulated reflection mainly comes from the surface reflection of the cover glass, and the intensity of this reflection is usually around 8% on an uncoated air-glass-air surface, and somewhere between 4% and 8% on an air-glass-LC surface [17,18]. Figure 5 considers three different reflection paths in the LCoS device as the example, with no modulation, normal modulation, and double modulation, respectively.

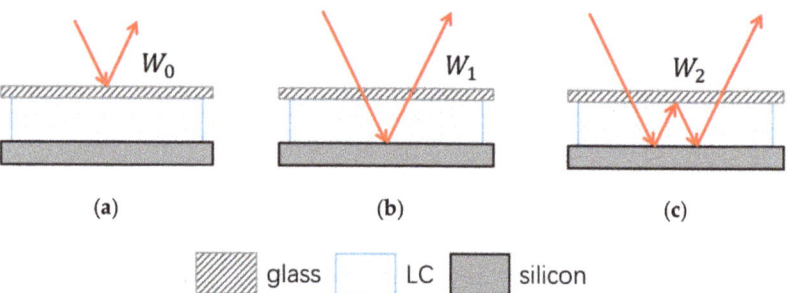

Figure 5. Different reflection paths in the LCoS device: (**a**) front reflection; (**b**) ideal condition; (**c**) back reflection. Glass thickness is shown in this figure but ignored in calculation.

The corresponding waveplates in each case are marked as W_0, W_1, and W_2, and their respective Jones matrices are shown as below. The mirroring effect caused by reflection is ignored since, in every case, the reflection count is an odd number.

$$W_0(\varphi_{eff}, \varphi_o) = \begin{bmatrix} 1 & 0 \\ 0 & 1 \end{bmatrix} \quad (9)$$

$$W_1(\varphi_{eff}, \varphi_o) = \begin{bmatrix} e^{i\varphi_{eff}} & 0 \\ 0 & e^{i\varphi_o} \end{bmatrix} \quad (10)$$

$$W_2(\varphi_{eff}, \varphi_o) = \begin{bmatrix} e^{i \cdot 2\varphi_{eff}} & 0 \\ 0 & e^{i \cdot 2\varphi_o} \end{bmatrix} \quad (11)$$

To model the glass reflection of the LCoS device, transmittance and reflectivity of the cover glass are defined. To simplify the model, the glass thickness is ignored, and reflections from all surfaces of the cover glass are treated as a single surface reflection, as the coherent length of laser in this experiment can be treated as infinite compared to the thickness of the cover glass. As a result, the single surface reflectivity is equal to the combined interference result of all the separated surfaces and its phase as a reference zero point. The glass absorption is ignored in all cases.

When the light is traveling from air toward LC (air-glass-LC), the transmittance of the cover glass is marked as A and the reflectivity as $(1 - A)$. With the same principle, when the light is traveling from LC toward air (LC-glass-air), the transmittance of the cover glass is marked as B and the reflectivity as $(1 - B)$. These two reflections are named front reflection (case W_0 in Figure 5) and back reflection (case W_2 in Figure 5) for simplicity. To simplify the expressions, A and B can be rewritten as:

$$A = cos^2\alpha, \quad 1 - A = sin^2\alpha \quad (12)$$

$$B = cos^2\beta, \quad 1 - B = sin^2\beta \quad (13)$$

These non-ideal reflections in W_0 and W_2 could cause systematic errors in the phase measurement of actual LCoS devices, with their principles illustrated in Appendix A. Such errors are further discussed with mathematical derivation and simulated experiments in subsequent subsections.

Simulations in the following sections are based on an 8-bit vertical-aligned (VA) LCoS device with an average LC layer thickness of 6 μm, and LC material with a refractive index of $n_o = 1.49$ and $n_e = 1.60$. Both front and back reflectivity were set to 5% unless specified otherwise, and a collimated 635 nm laser source with infinite coherence length was used. The non-uniform LC layer thickness was simulated by a spherical CMOS backplane with a curvature of 100 m, giving a peak–peak difference of 478 nm across its 12.3 mm × 6.9 mm active area. The thickness and non-uniformity of the LC layer were slightly amplified compared to a more common thickness of 3 μm and peak–peak difference of 100 nm [14] to emphasize the effect of glass reflection.

2.3. Theory and Analysis with Un-Modulated Front Reflection

In this subsection, only the front reflection of the cover glass is considered, which means that only W_0 and W_1 are included. W_2 is temporarily ignored in this subsection for simplicity reasons.

The Jones matrix of the LCoS device should be rewritten as:

$$W'(\varphi_{eff}, \varphi_o, \alpha) = cos\alpha \cdot W_1(\varphi_{eff}, \varphi_o) + sin\alpha \cdot W_0(\varphi_{eff}, \varphi_o)$$
$$= \begin{bmatrix} e^{i\varphi_{eff}} cos\alpha + sin\alpha & 0 \\ 0 & e^{i\varphi_o} cos\alpha + sin\alpha \end{bmatrix} \quad (14)$$

Further, the complex amplitude of the transmitted light can be calculated with

$$E'_{out} = P(\theta_2) W'\left(\varphi_{eff}, \varphi_o, \alpha\right) \sqrt{I_{in}} \begin{bmatrix} \cos\theta_1 \\ \sin\theta_1 \end{bmatrix} \tag{15}$$

The transmitted light intensity I'_{out} changes into the following form:

$$I'_{out} = I_{in} \cdot \left| \left(e^{i\Gamma} \cos\theta_1 \cos\theta_2 + \sin\theta_1 \sin\theta_2\right) e^{i\varphi_o} \cos\alpha + \cos(\theta_1 - \theta_2) \sin\alpha \right|^2 \tag{16}$$

Note the newly appeared term $\cos(\theta_1 - \theta_2) \cdot \sin\alpha$ on the right side. Apparently, either when the orthogonal relation between θ_1 and θ_2 is satisfied or the reflectivity $\sin^2\alpha$ equals to zero, the above expression would degenerate into the form obtained in Section 2.1, and the calculation should be error-free under this circumstance. This outcome is understandable, as the reflected component from W_0 has the same polarization state as polarizer P1 and should be filtered out by the second polarizer P2.

For example, when θ_1 and θ_2 are orthogonal to each other, and the transmittance of the cover glass $A = \cos^2\alpha$ is less than 1, I'_{out} can be expressed as follows:

$$I'_{out} = \frac{I_{in}}{2} \cos^2(\alpha) \cdot \sin^2(2\theta_1) \cdot (1 - \cos\Gamma) \tag{17}$$

However, in cases in which θ_1 and θ_2 are not orthogonal, and transmittance A is less than 1, the light intensity I'_{out} cannot be transformed into a simple form. In this case, the final output intensity is not only related to the phase retardation Γ but also related to the optical path difference φ_o, and it is no longer possible to calculate the phase retardation Γ with normalized output intensity I'_{norm}.

Considering that the reflectivity of the LCoS device is usually not controllable during the measurement, the proper alignment of the polarization orientation can be very important in this scenario. One possible source of such misalignment is the non-polarizing beam splitter (NPBS) used in this setup, which can bring up to 10° of polarization angle change and a certain amount of linearity degradation [19]. A possible solution to this problem is further discussed in Appendix A.

2.4. Theory and Analysis with Double-Modulated Back Reflection

In this case, only the back reflection of the cover glass is considered, which means only W_1 and W_2 are included, and W_0 is assumed to have been already corrected by the orthogonal polarizer pair, which means $\theta_1 - \theta_2 = \pm\frac{\pi}{2}$ is already satisfied, and light from the front reflection will not pass through the second polarizer P2. W_0 is ignored in the math derivations but still included in the simulations. Note that W_0 is ignorable only when P2 is orthogonal to P1 in this measuring process, which is not the case when LCoS is used in actual applications, such as holographic displays or wavelength selective switches, in which only extraordinary light is present [20,21].

The Jones matrix of the LCoS device should be rewritten as:

$$W''\left(\varphi_{eff}, \varphi_o, \beta\right) = \cos\beta \cdot W_1\left(\varphi_{eff}, \varphi_o\right) + \sin\beta \cdot W_2\left(\varphi_{eff}, \varphi_o\right)$$
$$= \begin{bmatrix} e^{i\varphi_{eff}} \cos\beta + e^{i \cdot 2\varphi_{eff}} \sin\beta & 0 \\ 0 & e^{i\varphi_o} \cos\beta + e^{i \cdot 2\varphi_o} \sin\beta \end{bmatrix} \tag{18}$$

Moreover, the transmitted light can be calculated as:

$$E''_{out} = P(\theta_2) W''\left(\varphi_{eff}, \varphi_o, \beta\right) \sqrt{I_{in}} \begin{bmatrix} \cos\theta_1 \\ \sin\theta_1 \end{bmatrix} \tag{19}$$

By applying the above assumed $\theta_1 - \theta_2 = \pm\frac{\pi}{2}$, the expressions for transmitted light intensity can be simplified into:

$$I''_{out} = \frac{1}{4}\sin^2(2\theta_1) \cdot I_{in} \cdot \left|\left(e^{i\Gamma} - 1\right)e^{i\varphi_o}\cos\beta + \left(e^{i\cdot 2\Gamma} - 1\right)e^{i\cdot 2\varphi_o}\sin\beta\right|^2 \quad (20)$$

As can be seen, only if the reflectivity $\sin^2\beta$ equals zero does the above expression for I''_{out} degenerate into the form of I_{out} in Equation (6), and it can be normalized into $I''_{norm} = (1 - \cos\Gamma)/2$. However, when the reflectivity cannot be ignored, $I''_{out} - \Gamma$ relation is additionally affected by the changing interference caused by double-modulation, and no longer follows the ideal sine-wave shape. Only when the phase retardation satisfies $\Gamma = 2k\pi, k \in Z$ can the distorted I''_{out} agree with its undistorted form I_{out} at its minimum value point. This finding indicates that only phase at $\Gamma = 2k\pi, k \in Z$ can be correctly measured. Note that, when $\Gamma = (2k+1)\pi, k \in Z$, although the form of I''_{out} agrees with I_{out}, it does not reach its maximum value due to the constructive interference next to this location, and the actual phase cannot be calculated at this point.

Figure 6a shows a simulation of the relation between I''_{out} and input gray level (GL) in one of the regions, while Figure 6b shows the change in average and peak–peak measurement error across the whole active area. An obvious relation between error and I''_{out} can be seen, that is, when I''_{out} reaches its minimum value, the measurement error also reaches its minimum, and the average phase can be viewed as error-free. This discovery is critical for the uniformity retrieving method discussed in the next subsection, which allows for uniformity calculation based on these error-free points at $\Gamma = 2\pi$.

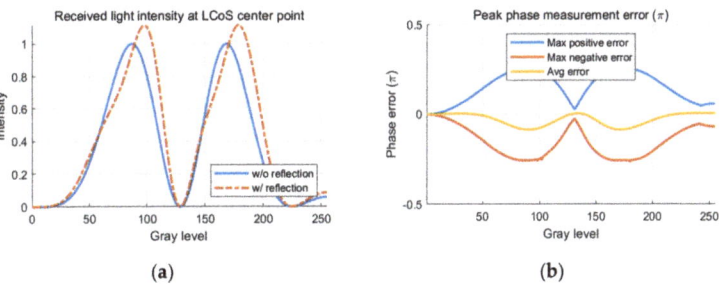

Figure 6. Simulation of the effect from 5% back reflection on I''_{out} and phase error: (**a**) $I''_{out} - GL$ relation; (**b**) peak and average phase error across the active area.

Note that the average measurement error across the active area is related to the LC layer thickness non-uniformity such that, when the thickness non-uniformity is greater, both positive error and negative error would appear and cancel each other out in the average value, while when the thickness non-uniformity is smaller, positive and negative errors are less likely to be balanced, and the average error would be greater and closer to the peak error.

2.5. Retrieving LCoS Uniformity from the Data Affected by Back Reflection

To address this error caused by back reflection and to obtain the correct uniformity data, the known error-free point at $\Gamma = 2k\pi$, $k \in Z$ can be utilized. Since the average phase result at this point can also be viewed as error-free, it is possible to retrieve the actual phase uniformity based on this set of data, assuming that the non-uniformity of phase response is solely determined by the non-uniformity of LC layer thickness. Additionally, in case the LC refractive indices $n_o(\lambda)$ and $n_e(\lambda)$ are known, and the maximum phase retardation can be reached at maximum GL, the uniformity of φ_o and the LC layer thickness could also be calculated. Figure 7 demonstrates the basic principle of this process with simulation data obtained in Section 2.4.

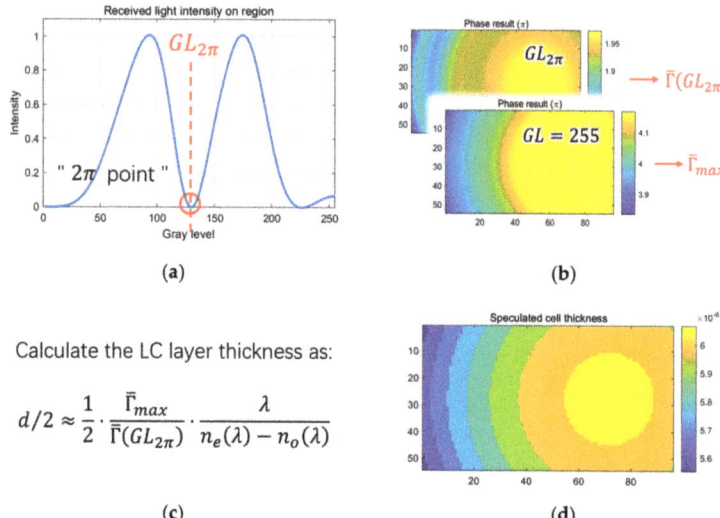

Figure 7. Basic principle of the LCoS uniformity extraction process: (**a**) locate the error-free point in the measured intensity curve of a specific region as $GL_{2\pi}$; (**b**) calculate the average phase of the whole active area at $GL_{2\pi}$ and max GL; (**c**) estimate the LC layer thickness in this region; (**d**) repeat for all regions and acquire the uniformity result.

For each measured region, we first find the designated low point where I''_{out} reaches its minimum value, and the phase retardation is 2π. We mark the gray level at this point as $GL_{2\pi}$. This low point should be the closest one to the zero phase retardation point, which should be the first low point after zero driving voltage for a common vertical-aligned LCoS (VA-LCoS) and the last low point before max driving voltage for common parallel-aligned LCoS (PA-LCoS) [22].

Then, we find the average measured phase of the whole active area at the max phase retardation point as $\overline{\Gamma}_{max}$, and the average measured phase across the whole active area at $GL_{2\pi}$ as $\overline{\Gamma}(GL_{2\pi})$. Note that both $\overline{\Gamma}_{max}$ and $\overline{\Gamma}(GL_{2\pi})$ are calculated from the original faulty result. The full phase retardation in this region can be approximately calculated as follows, assuming that the average error of the phase measurement at $GL_{2\pi}$ is much smaller than the peak error, as demonstrated in Figure 6b.

$$\Gamma_{max} = 2\pi \cdot \overline{\Gamma}_{max}/\overline{\Gamma}(GL_{2\pi}) \tag{21}$$

Since we have assumed that the phase retardation is only related to LC layer thickness, the thickness can be calculated by multiplying phase retardation by $\lambda/(2\pi \cdot n_o(\lambda))$, and the final expression of LC layer thickness $d/2$ at this location can be written as:

$$d/2 \approx \frac{1}{2} \cdot \frac{\overline{\Gamma}_{max}}{\overline{\Gamma}(GL_{2\pi})} \cdot \frac{\lambda}{n_e(\lambda) - n_o(\lambda)} \tag{22}$$

Repeating such processes for all regions of the LCoS active area, we can then acquire a uniformity map of the LC layer thickness on this LCoS device.

3. Results

3.1. Results of the Front Reflection

According to the theory and modeling in Section 2.3, a simulated measurement when the second polarizer P2 is misaligned by 10 degrees is calculated, as demonstrated in Figure 8. Ripple-shaped measurement error can be seen in the result.

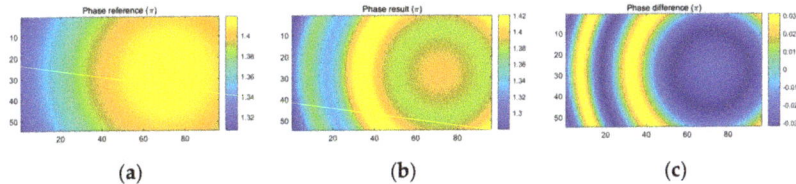

Figure 8. Simulated result when the polarizers are misaligned by 10 degrees: (**a**) actual phase in π; (**b**) measured phase in π; (**c**) measurement error in π.

Figure 9 shows the result of a measured phase uniformity on an LCoS device. The polarization orientation was properly aligned in Figure 9a, while Figure 9b presents the result when the orientation was misaligned by 8 degrees due to NPBS distortion. Similar patterns could be seen compared to the simulated result in Figure 8, and the amplitude of measurement error was also in line with the simulation. However, since the amplitude of this error caused by the front reflection was very small (usually less than $\pm 0.02\pi$), it could easily be confused with other noises in the measurement process, especially when the phase flicker was large. The conclusion by Martínez et al. in their demonstration of the classical polarimetric method, stating that high quality NPBS does not present a considerable difference in LCoS phase measurement [11], can be confirmed according to our simulations and experiments.

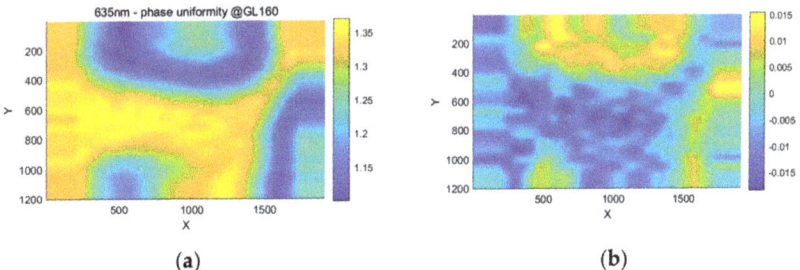

Figure 9. Real-life LCoS phase measurement: (**a**) measured phase in π with correct alignment; (**b**) measurement error in π, with 8° of P2 misalignment.

3.2. Results of the Back Reflection

As discussed and analyzed in Section 2.4, the results when back reflection is present should be discussed in two separate parts, with one part focusing the gray level where maximum error occurs, and another part focusing the gray level where only minimum error occurs.

Figure 10 demonstrates the simulated worst-case scenario, when I''_{out} and the measurement error are at the maximum, as indicated in Figure 10a. Compared to the real phase in Figure 10b, the measured phase distribution in Figure 10c shows a clear ripple-shaped error pattern, and the phase error can reach up to $\pm 0.25\pi$, as demonstrated in Figure 10d, completely invalidating the measurement result. The error pattern shows a clear periodical relation with the LC layer thickness, resulting in a rippled shape.

Notice that the max amplitude of such a phase measurement error has a positive relationship with the reflectivity of the cover glass. According to the simulation, when the reflectivity is 1%, the maximum measurement error is around $\pm 0.12\pi$, which is lower than that from a 5% reflection.

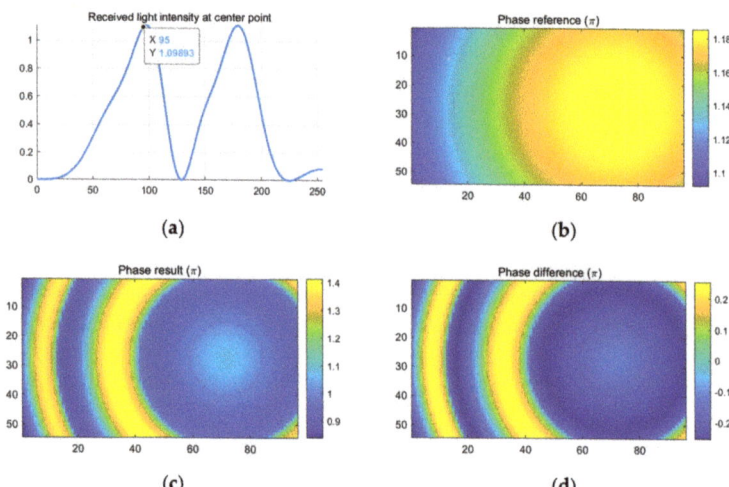

Figure 10. Simulated result with 5% back reflection (maximum error point): (**a**) location of data point; (**b**) actual phase in π; (**c**) measured phase in π; (**d**) measurement error in π.

Meanwhile, Figure 11 demonstrates the simulated best-case scenario when I''_{out} and the measurement error are at the minimum, as indicated in Figure 11a. Compared to the real phase in Figure 11b, the measured phase distribution in Figure 11c shows only very little measurement error, which is around $\pm 0.025\pi$.

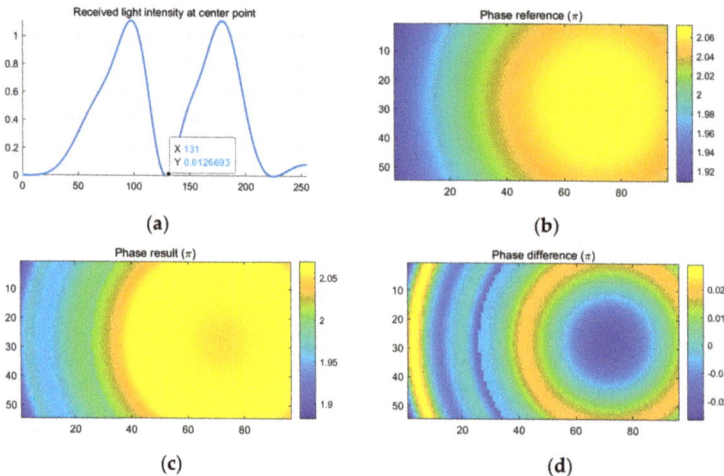

Figure 11. Simulated result with 5% back reflection (minimum error point): (**a**) location of data point; (**b**) actual phase in π; (**c**) measured phase in π; (**d**) measurement error in π.

The phase error across the whole active area never reaches zero, as shown in Figure 6b, since the non-uniformity of the LC layer indicates that different areas would reach their error-free point of $\Gamma = 2k\pi, k \in Z$ at different input gray levels, and there will always be some area with measurement error regardless of the gray level used.

The effect of the back reflection in an actual real-life measurement is compared to the simulation and analyzed in subsequent paragraphs. According to the discussion in Section 2.4, when the measured light intensity is around its maximum value, the phase measurement error also reaches its maximum. Figure 12 demonstrates the measured intensity-GL relationship (Figure 12a) and the calculated phase distribution (Figure 12) at the indicated GL127 input. As shown in the figure, the amplitude of this ripple-shaped pattern is about $\pm 0.15\pi$, in line with the simulated results in Section 2.4 and Figure 10. This error cannot be ignored in the phase uniformity measurement.

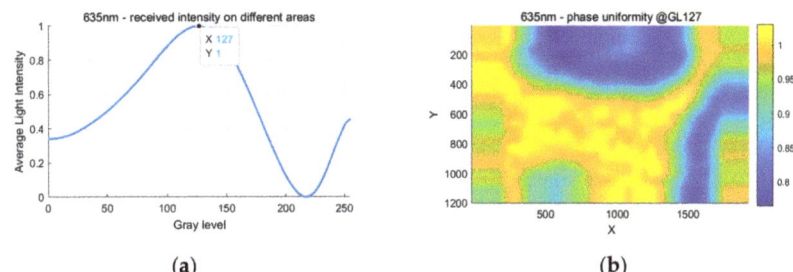

Figure 12. Phase measurement at near $(2k+1)\pi$ phase: (**a**) intensity-GL curve and indication of the data point; (**b**) calculated phase distribution in π.

Figure 13 demonstrates the measured intensity–GL relationship (Figure 13a) and the calculated phase distribution (Figure 13b) at the indicated GL212 input. The amplitude of the error pattern is only around $\pm 0.02\pi$ in this condition, which also agrees with the simulated result in Figure 11.

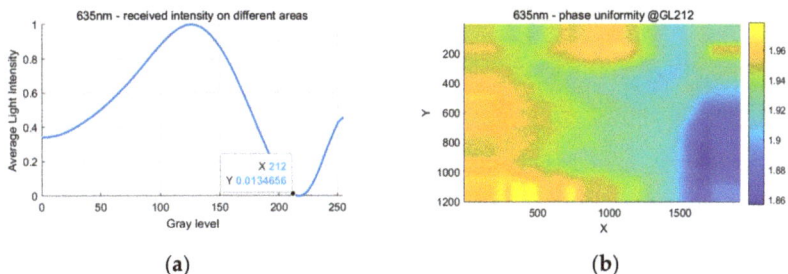

Figure 13. Phase measurement at near $2k\pi$ phase: (**a**) intensity-GL curve and indication of the data point; (**b**) calculated phase distribution in π.

By comparing the shape and amplitude of such ripple-shaped patterns under different GL inputs, the results of our experiment match the theory in Section 2 quite well, indicating that the theory is effective and can be used to guide the LCoS uniformity retrieving process.

3.3. Results of the LCoS Uniformity Retrieving

Figure 14 demonstrates a simulation of our uniformity retrieving process, assuming the birefringence refractive index is accurate, and the max phase retardation is reached at $GL = 255$. Figure 14a shows the speculated LC layer thickness based on measured light intensity I''_{out}, and the error of such speculation is less than 1%, as shown in Figure 14b.

One main source of the error is the discrete gray level control, which makes it impossible to locate the exact low point between two neighboring gray levels, where I''_{out} reaches its minimum. The other main source of error is the difference between the average measured phase and the actual phase, as demonstrated in Figure 6b. Additionally, the discrete gray level addressed also makes the resolution of this uniformity result relatively low, as can be seen in Figure 14a with discrete steps.

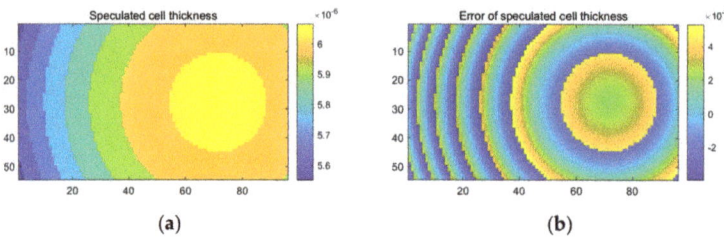

Figure 14. Speculated LC layer thickness in simulation: (**a**) speculated thickness; (**b**) error of the speculation.

This simulation indicates that it is actually a viable option for roughly measuring the uniformity of LC layer thickness on LCoS devices, with the drawback of having to measure the full modulation range of the LCoS device and having to know the exact refractive indices of the LC material. However, if the absolute value of the thickness is not needed, both the full modulation range and the exact LC refractive indices are no longer required. The relative uniformity can still be obtained by only measuring a single full 2π phase range of the whole active area. In this case, only the relative relation in the calculated uniformity map holds true.

For the actual measured data in real-life experiments, the LC thickness uniformity can be obtained based on Equation (22). Figure 15a shows the calculated LC layer thickness of this LCoS device indexed by subregions, and a clear saddle-shaped pattern can be seen. By comparing it to the interference pattern obtained on the same LCoS device shown in Figure 15b, obvious similarities can be seen, and the calculated thickness difference also matches the number of interference stripes, indicating the effectiveness of this method, which extracts the uniformity of the LCoS device.

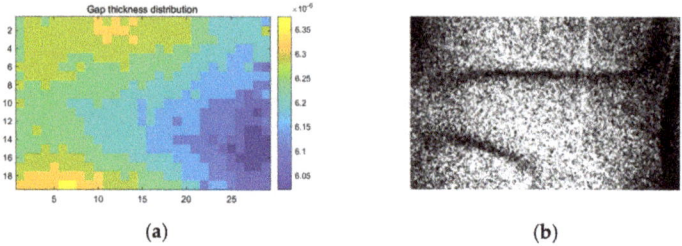

Figure 15. Result of the correction: (**a**) calculated LC layer thickness; (**b**) interference pattern.

The pattern and peak–peak difference in this calculated LC layer thickness also agrees with the common saddle-shaped cell gap thickness pattern on LCoS devices demonstrated in other research groups' work [14], indicating that this process is indeed effective at phase around $2k\pi$, $k \in Z$.

It is also worth mentioning that, if the LC refractive index is inaccurate, or the max phase retardation is not reached, the absolute value of this thickness speculation will also be inaccurate as a linear scale of the real value. However, even in this case, the relative uniformity is still preserved.

4. Discussion

The LC thickness uniformity of our LCoS device has been calculated in Section 3, but the linearity of the phase response $\varGamma$ is still not fully obtained. According to Equation (20), if φ_0 is accurately measured, the phase retardation $\varGamma$ can be calculated. However, in our simulation setup, with an average LC layer thickness of 6 μm and $n_o = 1.49$ at 635 nm, the average value of the total ordinary phase φ_0 is around 55π, which means the accuracy of measured φ_o is only about 0.5π, not including other possible errors caused by inaccurate refractive indices or driving strength. Such accuracy is not enough to calculate phase retardation Γ based on I''_{out} in Equation (20), which requires $Err(\varphi_o) \ll \pi$. Therefore, an accurate result of Γ was not yet achievable in this setup.

On the other hand, since the relative uniformity of the LC layer thickness is known, the uniformity of phase retardation can still be hypothesized based on this calculated data, assuming that the phase response is only determined by the LC layer thickness and driving voltage, ignoring all other possible non-uniform aspects.

First, calculate the phase of reflected ordinary light based on the measured LC layer thickness $d/2$. The uniformity distribution of the property (e.g., $\boldsymbol{\varphi}$, $\boldsymbol{d}$ or $\boldsymbol{\Gamma}$) is marked in bold to differentiate from that the measured value from a single point. The result in this step might not be accurate, but it should be proportional to the real value.

$$\varphi_o = \frac{2\pi \cdot n_o(\lambda) \cdot \boldsymbol{d}}{\lambda} \qquad (23)$$

Then, for each gray level GL, assume the distribution of phase retardation $\boldsymbol{\Gamma}$ is proportion to the distribution of ordinary phase difference $\boldsymbol{\varphi_o}$, which means $\boldsymbol{\Gamma} = k \cdot \boldsymbol{\varphi_o}$. The least square method can be used to determine the coefficient value k, causing the speculated Γ to match the measured Γ'' in the experiment. Repeating this process for all gray levels, the approximate phase distribution can be acquired. Figure 16 shows the result of such a process with simulation data.

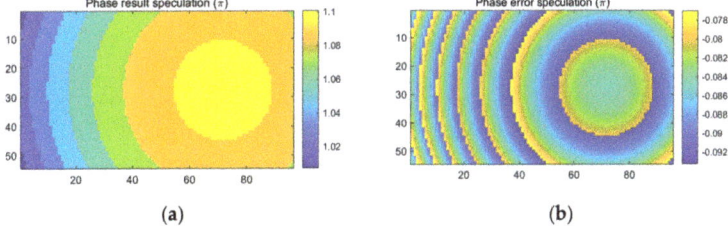

Figure 16. Phase speculation based on the LC layer thickness uniformity result: (**a**) speculated phase at $GL = 95$ where error reaches maximum; (**b**) error of the speculation at $GL = 95$.

Such speculation is far from perfect. As shown in Figure 16b, error with this speculation method is still determined by the average phase error shown in Figure 6b, which reaches up to 0.09π. Since the least square fitting is used in the speculation process, the average error of the new speculated phase and the original faulty phase is close to the same, about -0.08π at this gray level. On the other hand, the root mean square of error (RMSE) in this case becomes much lower at only 43% of the original RMSE, which is a huge improvement. The error comparison can be seen in Table 1.

Table 1. Simulated comparison between the original polarimetric method and our new method at $GL = 95$, where the error reaches its maximum.

Phase Calculation Methods	Average Error (π)	RMSE (π)
Original polarimetric method [10]	-8.36×10^{-2}	1.97×10^{-1}
Our method	-8.53×10^{-2}	8.54×10^{-2}

Apart from the classical polarimetric method discussed in this paper, another outcome is also worth noting. The effect of back reflection is mainly due to the changing interference caused by overmodulation under different phase inputs, so it is not limited to uniformity measurements and not only affects the classical polarimetric method discussed in this paper but can also affect other LCoS phase measuring methods based on phase retrieving from normalized light intensity, including the binary grating method [23] and some interference methods [4,5] to a certain extent. Figure 17 shows the measured phase result with binary grating methods on the same LCoS device, and a similar ripple-shaped error pattern is also present in this result.

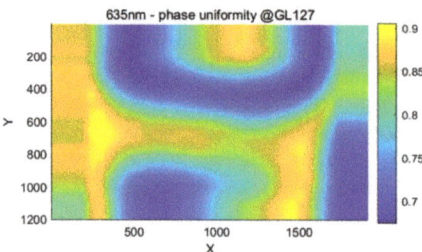

Figure 17. Measured phase in π with binary grating method.

One possible way to suppress the effect of back reflection is by using a light source with a very short coherence length, such as an incandescent light source with a coherence length less than the LC layer thickness [24]. In this way, interference will not happen on the glass surface, which is thought to greatly reduce the amplitude of error caused in the measurement result. Additional research could be conducted by comparing the difference between coherent and non-coherent light sources.

5. Conclusions

In this paper, the effect of glass reflections on LCoS phase uniformity measurement was discussed under three conditions. Based on the idealized modeling of the classical polarimetric method for LCoS phase measurement, a more detailed analytical model, including front and back reflections from its cover glass, was constructed, and the simulation result based on this new model was compared to real-life measurements with similar parameters. According to the results comparison, the newly proposed model with included glass reflections can explain the distinct ripple-shaped pattern in the phase measurement result.

In the case of front reflection, where light was reflected before being modulated by the LC layer, the error caused by reflection was easily corrected by removing the NPBS and realigning the polarizers. Even if the error is unattended, its amplitude is maxed out around $\pm 0.02\pi$ with the non-AR-coated cover glass.

In the case of back reflection, when light was reflected back into the LC layer by the cover glass, the error reached an amplitude of $\pm 0.2\pi$ or more, causing the uniformity measurement to be completely ineffective. Furthermore, when the calibration process was performed based on this faulty measurement result, the actual phase distribution became more uneven. This phenomenon is thought to be one of the reasons that this non-ideal glass reflection was overlooked in most previous works.

A method to acquire the approximate phase distribution was proposed by calculating the uniformity of LC layer thickness based on a full-range measurement of the LCoS device. The phase uniformity was estimated with sub-0.1π accuracy and half RMSE compared to the original result, assuming a linear relationship between the phase value and LC layer thickness. Further research is needed, with investigations of the multipath reflections of LCoS devices, as well as the effect of other non-uniform aspects, such as LCoS flatness and electric field strength, to make the classical polarimetric method a viable and consistent means of measuring LCoS phase uniformity.

Author Contributions: Conceptualization, X.Z. and K.L.; methodology, X.Z. and K.L.; software, X.Z.; validation, X.Z. and K.L.; formal analysis, X.Z.; investigation, X.Z.; resources, K.L.; data curation, X.Z.; writing—original draft preparation, X.Z.; writing—review and editing, K.L.; visualization, X.Z.; supervision, K.L.; project administration, K.L.; funding acquisition, K.L. All authors have read and agreed to the published version of the manuscript.

Funding: This research was supported by the Fundamental Research Funds for the Central Universities, project no. 2242023K40002, and by the Start-up Research Fund of Southeast University, project no. RF1028623165.

Data Availability Statement: The data presented in this study are available in the article.

Acknowledgments: The authors would like to acknowledge the support from CamOptics (Suzhou) Ltd., for supplying the LCoS device used in some of the experiments.

Conflicts of Interest: The authors declare no conflict of interest.

Appendix A

The effect of non-ideal reflections on the actual phase retardation is analyzed in Appendix A. The phase retardation in the output extraordinary light with front and back reflections included can be written as:

$$\begin{aligned}\Gamma' &= \text{angle}\left(\sqrt{1-A} \cdot e^{i \cdot 0} + \sqrt{A \cdot B} \cdot e^{i \cdot \Gamma} + \sqrt{A \cdot (1-B)} \cdot e^{i \cdot 2\Gamma}\right) \\ &= \text{angle}\left(\sin\alpha \cdot e^{i \cdot 0} + \cos\alpha \cdot \cos\beta \cdot e^{i \cdot \Gamma} + \cos\alpha \cdot \sin\beta \cdot e^{i \cdot 2\Gamma}\right)\end{aligned} \quad (A1)$$

When the front and back reflectivity satisfies $(1 - A) = (1 - B)$, and $(1 - A)$ and $(1 - B)$ are small enough, Γ' approximately equals to Γ. In other words, it can be seen that the reflections do not change the actual phase retardation on the LCoS device. With some calculation, it can be found out that, when $(1 - A) = (1 - B) \leq 0.11$, the change of phase retardation $|\Gamma' - \Gamma|$ is less than 0.01π, which can be ignored in simulations and discussions within this article.

The components of the reflected light are plotted to illustrate the effect of glass reflections on the output light intensity. Figure A1 shows a simulation of reflected light before and after the second polarizer P2, separated as individual components. Assume $\theta_1 = -\frac{\pi}{4}$ and $\theta_2 = \frac{\pi}{4}$, respectively, and both glass reflectivities $(1 - A) = (1 - B) = 0.05$. Input intensity is normalized to 1 for both ordinary light and extraordinary light. Take $\varphi_o = 50\pi$ and $\Gamma = 0.6\pi$ in this simulation.

Figure A1a demonstrates the extraordinary light reflections separated as three components from W_0, W_1, and W_2. As can be seen in the plot, the amplitude of the combined reflection has been changed due to the existence of W_0 and W_2, while the phase remains unchanged compared to the reflection from W_1. With the same principle, Figure A1b demonstrates the ordinary reflections, which also have their amplitude changed but their phase unchanged.

Finally, Figure A1c demonstrates the combined light after polarizer P2, with the solid lines representing the output light with non-ideal reflections and their dashed counterparts representing the output light under ideal conditions. As can be seen, the amplitude of the combined light has also changed due to the amplitude change in both ordinary and extraordinary light. Change in the output amplitude is related to φ_o and Γ, making the final output intensity impossible to be normalized into the ideal form of $I_{norm} = (1 - \cos\Gamma)/2$. This outcome will result in the ripple-shaped error pattern demonstrated in the introduction section.

Figure A2 demonstrates a common optical setup for polarimetric method measurement. A single collimated laser beam went through polarizer P1 and NPBS and then was shaped by an aperture and reflected off a certain region of the LCoS device. The beam was steered off the main optical path by the NPBS, went through polarizer P2, and was collected by the photo detector.

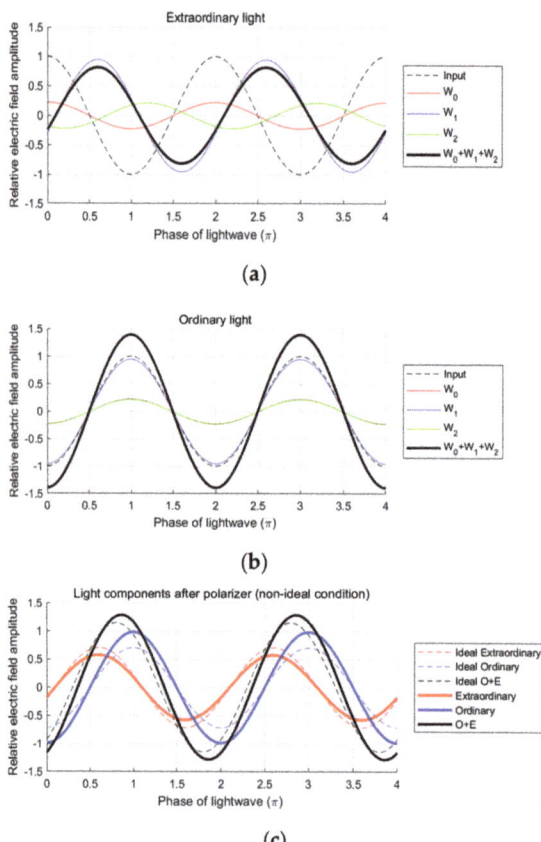

Figure A1. Comparison of the light components before and after polarizer P2: (**a**) extraordinary reflections before P2; (**b**) ordinary reflections before P2; (**c**) combined reflections after P2.

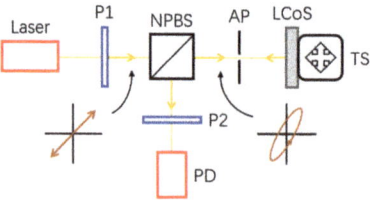

Figure A2. Non-ideal optical setup for polarimetric method (P, polarizer, NPBS, non-polarizing beam splitter, AP, aperture, TS, translation stage, PD, photo detector).

However, when the incident light is not fully S-polarized or P-polarized, which corresponds to X-axis or Y-axis linear polarization, respectively, in this setup, NPBS cannot guarantee the consistency between the incident light polarization state and the transmitted light polarization state, and it could cause a change in polarization orientation and a degraded polarization linearity. According to testing conducted by Thorlabs, Inc., even with the best-case scenario, the change of polarization angle reaches up to 5 degrees, not including the degraded linearity [19]. Unfortunately, the incident ± 45° linearly polarized light and the reflected elliptical polarized light meet this exact condition, indicating the NPBS is likely to cause systematic error during the measurements.

Considering the uncertain distortion from the NPBS, a quasi-perpendicular setup is preferred in the measuring process. After passing through polarizer P1 and aperture AP1, the laser beam is steered directly by the LCoS device through polarizer P2 and aperture AP2, and it is collected by the photo detector. Such a modified setup is demonstrated in Figure A3.

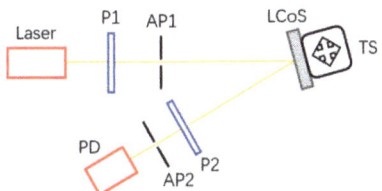

Figure A3. Improved optical setup for polarimetric method (P, polarizer, AP, aperture, TS, translation stage, PD, photo detector).

To ensure the alignment between two polarizers, it is recommended to replace the LCoS device with a reflection mirror before the measurement, so the laser beam can be directly reflected into polarizer P2 after polarizer P1 without phase modulation between them. Adjust the transmissive direction of P2 until the received light intensity on the photo detector reaches the minimum. The polarizers should have been properly aligned to an orthogonal state after the adjustment, and the measurement can be started after switching the mirror back to the LCoS device under testing.

References

1. Lazarev, G.; Chen, P.-J.; Strauss, J.; Fontaine, N.; Forbes, A. Beyond the Display: Phase-Only Liquid Crystal on Silicon Devices and Their Applications in Photonics [Invited]. *Opt. Express* **2019**, *27*, 16206–16249. [CrossRef] [PubMed]
2. Chen, H.-M.P.; Yang, J.-P.; Yen, H.-T.; Hsu, Z.-N.; Huang, Y.; Wu, S.-T. Pursuing High Quality Phase-Only Liquid Crystal on Silicon (LCoS) Devices. *Appl. Sci.* **2018**, *8*, 2323. [CrossRef]
3. Tong, Y.; Pivnenko, M.; Chu, D. Improvements of Phase Linearity and Phase Flicker of Phase-Only LCoS Devices for Holographic Applications. *Appl. Opt.* **2019**, *58*, G248–G255. [CrossRef] [PubMed]
4. Liebmann, M.; Valverde, J.; Kerbstadt, F. Wavefront Compensation for Spatial Light Modulators Based on Twyman-Green Interferometry. In Proceedings of the Advances in Display Technologies XI, Bellingham, WA, USA, 5 March 2021; Volume 11708, pp. 142–149.
5. Zeng, Z.; Li, Z.; Fang, F.; Zhang, X. Phase Compensation of the Non-Uniformity of the Liquid Crystal on Silicon Spatial Light Modulator at Pixel Level. *Sensors* **2021**, *21*, 967. [CrossRef] [PubMed]
6. Otón, J.; Ambs, P.; Millán, M.S.; Pérez-Cabré, E. Multipoint Phase Calibration for Improved Compensation of Inherent Wavefront Distortion in Parallel Aligned Liquid Crystal on Silicon Displays. *Appl. Opt.* **2007**, *46*, 5667–5679. [CrossRef]
7. Bing, Z.; Yuntian, T.; Lili, X.; Qiong, W. The Vibration Isolation Technologies of Load in Aviation and Navigation. *Int. J. Multimed. Ubiquitous Eng.* **2015**, *10*, 19–26. [CrossRef]
8. Gong, W.; Li, A.; Huang, C.; Che, H.; Feng, C.; Qin, F. Effects and Prospects of the Vibration Isolation Methods for an Atomic Interference Gravimeter. *Sensors* **2022**, *22*, 583. [CrossRef]
9. Charrière, F.; Kühn, J.; Colomb, T.; Montfort, F.; Cuche, E.; Emery, Y.; Weible, K.; Marquet, P.; Depeursinge, C. Characterization of Microlenses by Digital Holographic Microscopy. *Appl. Opt.* **2006**, *45*, 829–835. [CrossRef]
10. Márquez, A.; Martínez, F.J.; Gallego, S.; Ortuño, M.; Francés, J.; Beléndez, A.; Pascual, I. Classical Polarimetric Method Revisited to Analyse the Modulation Capabilities of Parallel Aligned Liquid Crystal on Silicon Displays. In Proceedings of the Optics and Photonics for Information Processing VI, Bellingham, WA, USA, 15 October 2012; Volume 8498, pp. 179–189.
11. Martínez, F.J.; Márquez, A.; Gallego, S.; Frances, J.; Pascual, I. Extended Linear Polarimeter to Measure Retardance and Flicker: Application to Liquid Crystal on Silicon Devices in Two Working Geometries. *OE* **2014**, *53*, 014105. [CrossRef]
12. Yang, Z.; Wu, S.; Nie, J.; Yang, H. Uncertainty in the Phase Flicker Measurement for the Liquid Crystal on Silicon Devices. *Photonics* **2021**, *8*, 307. [CrossRef]
13. Zhang, Z.; Yang, H.; Robertson, B.; Redmond, M.; Pivnenko, M.; Collings, N.; Crossland, W.A.; Chu, D. Diffraction Based Phase Compensation Method for Phase-Only Liquid Crystal on Silicon Devices in Operation. *Appl. Opt.* **2012**, *51*, 3837–3846. [CrossRef]
14. Zhang, Z.; Jeziorska-Chapman, A.M.; Collings, N.; Pivnenko, M.; Moore, J.; Crossland, B.; Chu, D.P.; Milne, B. High Quality Assembly of Phase-Only Liquid Crystal on Silicon (LCOS) Devices. *J. Display Technol.* **2011**, *7*, 120–126. [CrossRef]
15. Van Gelder, R.; Melnik, G. Uniformity Metrology in Ultra-Thin LCoS LCDs. *J. Soc. Inf. Disp.* **2006**, *14*, 233–239. [CrossRef]

16. Ronzitti, E.; Guillon, M.; Sars, V.; Emiliani, V. LCoS Nematic SLM Characterization and Modeling for Diffraction Efficiency Optimization, Zero and Ghost Orders Suppression. *Opt. Express* **2012**, *20*, 17843–17855. [CrossRef]
17. Dey, T.; Naughton, D. Cleaning and Anti-Reflective (AR) Hydrophobic Coating of Glass Surface: A Review from Materials Science Perspective. *J. Sol-Gel. Sci. Technol.* **2016**, *77*, 1–27. [CrossRef]
18. Prado, R.; Beobide, G.; Marcaide, A.; Goikoetxea, J.; Aranzabe, A. Development of Multifunctional Sol–Gel Coatings: Anti-Reflection Coatings with Enhanced Self-Cleaning Capacity. *Sol. Energy Mater. Sol. Cells* **2010**, *94*, 1081–1088. [CrossRef]
19. Thorlabs, Inc. Output Optical Properties of Beamsplitters with Angle of Incidence. Available online: https://www.thorlabs.com/images/TabImages/Beamsplitter_Lab.pdf (accessed on 17 March 2023).
20. Han, Z.; Yan, B.; Qi, Y.; Wang, Y.; Wang, Y. Color Holographic Display Using Single Chip LCOS. *Appl. Opt.* **2019**, *58*, 69–75. [CrossRef]
21. Wang, M.; Zong, L.; Mao, L.; Marquez, A.; Ye, Y.; Zhao, H.; Vaquero Caballero, F.J. LCoS SLM Study and Its Application in Wavelength Selective Switch. *Photonics* **2017**, *4*, 22. [CrossRef]
22. Zhang, Z.; You, Z.; Chu, D. Fundamentals of Phase-Only Liquid Crystal on Silicon (LCOS) Devices. *Light Sci. Appl.* **2014**, *3*, e213. [CrossRef]
23. Yang, S.; Yang, H.; Qin, L.; Shi, Y.; Li, Q. Measuring the Relationship between Grayscale and Phase Retardation of LCoS Based on Binary Optics. *SID Symp. Dig. Tech. Pap.* **2020**, *51*, 140–143. [CrossRef]
24. Donges, A. The Coherence Length of Black-Body Radiation. *Eur. J. Phys.* **1998**, *19*, 245. [CrossRef]

Disclaimer/Publisher's Note: The statements, opinions and data contained in all publications are solely those of the individual author(s) and contributor(s) and not of MDPI and/or the editor(s). MDPI and/or the editor(s) disclaim responsibility for any injury to people or property resulting from any ideas, methods, instructions or products referred to in the content.

Article

Design of Tunable Liquid Crystal Lenses with a Parabolic Phase Profile

Wenbin Feng, Zhiqiang Liu, Hao Liu and Mao Ye *

School of Optoelectronic Science and Engineering, University of Electronic Science and Technology of China, Chengdu 611730, China
* Correspondence: mao_ye@uestc.edu.cn

Abstract: An electrode pattern design generating a parabolic voltage distribution, in combination with usage of the linear response range of the liquid crystal (LC) material, has been recently proposed to obtain nearly ideal phase profiles for LC lenses. This technique features low driving voltages, simple structure, compact design, and the absence of high-resistivity (HR) layers. In this work, the universal design principle is discussed in detail, which is applicable not only to LC lens design, but also to other LC devices with any phase profile. Several electrode patterns are presented to form a parabolic voltage distribution. An equivalent electric circuit of the LC lens based on the design principle is developed, and the simulation results are given. In the experiments, an LC lens using the feasible parameters is prepared, and its high-quality performance is demonstrated.

Keywords: design method; LC lens; parabolic profile; circuit

1. Introduction

Liquid crystal (LC) is the state of aggregation that possesses the properties of both a crystalline solid and an isotropic liquid [1], which has been extensively researched [2–5]. The molecular orientation of LC can be controlled with applied electric/magnetic/optical fields leading to the tunable electrical, electro-optical and optical properties [5], allowing for LC to be used in many devices, such as LC displays and LC lenses. The LC lens was first proposed by Sato in 1979 [6] and has been attracting much attention in recent decades [7–18]. The focal length of the LC lens is electrically tunable compared with conventional solid lenses. The LC lenses using hole-patterned electrodes have been studied extensively [19–25], but the substrate between the patterned electrode and the LC layer causes high driving voltages. High-resistivity (HR) layers make the use of the substrate unnecessary, thus making low voltages possible [26–30]. The resistance of the HR layer usually changes over time, which results in the properties of the LC lens becoming unstable. Some other structures, including the use of transmission lines [31–37], surf relief electrodes [38,39] and lens-shaped dielectric material [40–42], have been reported. It is difficult to obtain LC lenses with a parabolic phase profile using these techniques. Multielectrode design makes nearly ideal lens properties possible [43–45]; however, too many voltage sources are required to drive the lens [46].

Recently, we proposed [47–50] forming parabolic voltage distributions using resistive lines and driving the LC lens with voltages within the linear response range of the LC material. An LC lens with a nearly parabolic phase profile is realized and the parabolic phase profile is maintained during focus tuning. In this work, we report several resistive line patterns and discuss the universal design principle in detail. The voltage distribution is simulated, and an LC lens with an aperture of 2 mm is demonstrated. The experimental and simulation results are in agreement.

Citation: Feng, W.; Liu, Z.; Liu, H.; Ye, M. Design of Tunable Liquid Crystal Lenses with a Parabolic Phase Profile. *Crystals* **2023**, *13*, 8. https://doi.org/10.3390/cryst13010008

Academic Editors: Zhenghong He and Yuriy Garbovskiy

Received: 1 December 2022
Revised: 16 December 2022
Accepted: 17 December 2022
Published: 21 December 2022

Copyright: © 2022 by the authors. Licensee MDPI, Basel, Switzerland. This article is an open access article distributed under the terms and conditions of the Creative Commons Attribution (CC BY) license (https://creativecommons.org/licenses/by/4.0/).

2. Principle

As shown in Figure 1, suppose a uniform resistive line is applied with voltages V_1 and V_2 at both ends. The voltage values at $N+1$ points on the line are taken and are rearranged at equal intervals to generate the desired voltage distributions determined by the length of the resistive line between these points. For example, if these points are equally spaced, the voltage distribution after rearrangement is linear. In particular, when the length of the resistive line between these points increases linearly, according to Ohm's law, the voltage value at each point is

$$V(n) = V_1 + \frac{\sum_{i=1}^{n}(i-1)l}{\sum_{i=1}^{N+1}(i-1)l}(V_2 - V_1) = V_1 + \frac{n^2}{N(N+1)}(V_2 - V_1) \quad (n = 1, 2, 3, \cdots, N+1) \tag{1}$$

where l is the length of the resistive line between the first two points. From Equation (1), we find that the voltage distribution is a parabolic function of point index n. After rearranging these points equidistantly along the x-axis, the voltage distribution is the parabolic function of x.

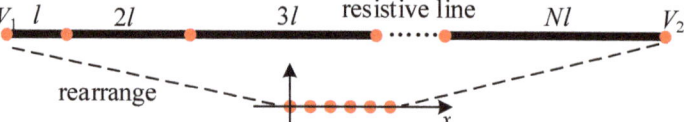

Figure 1. The schematic diagram of a resistive line.

In fact, a parabolic voltage distribution can be obtained without considering the shape of the resistive line as long as the length of the resistive line between these points increases linearly from zero. Thus, we can design the shape of the resistive line to make the electrode more compact and avoid the rearrangement process. Several examples are shown in Figure 2; they form a parabolic voltage distribution along the x-axis.

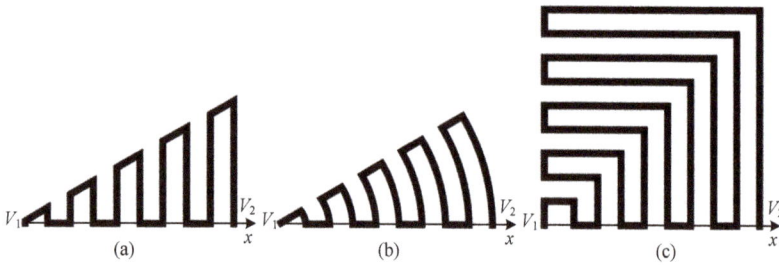

Figure 2. Several electrode patterns to form a parabolic voltage distribution. (**a**) triangle, (**b**) fan-shaped, and (**c**) rectangle.

The electrode structures shown in Figure 2 are called a "generating unit". The parabolic voltage distribution of a generating unit should be distributed over the lens aperture by a "distributing unit". For a cylindrical LC lens, two identical generating units should be arranged symmetrically to form a symmetrical parabolic voltage distribution, which is then distributed over the lens aperture by a series of parallel ITO electrodes. Taking the first two structures in Figure 2 as examples, the complete electrodes are shown in Figure 3. The solid-filled area represents the resistive line of the generating unit, and the slash-filled area represents the conductive line of the distributing unit.

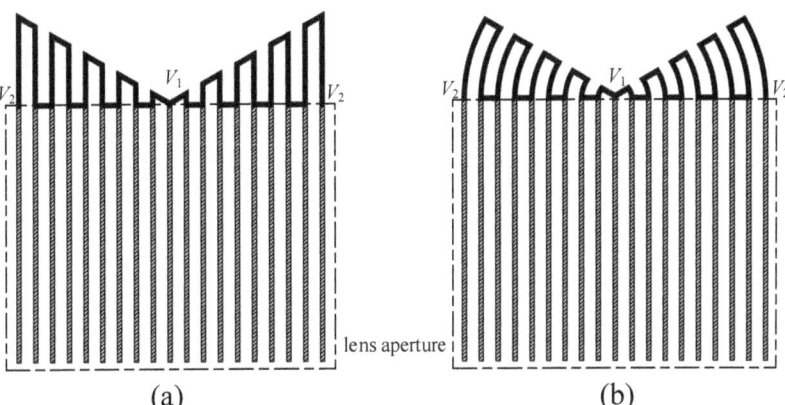

Figure 3. The electrode patterns of cylindrical LC lenses based on different generating units. (**a**) triangle type and (**b**) fan-shaped type.

The fan-shaped generating unit shown in Figure 2b can also be used for circular LC lenses, as shown in Figure 4. The first way is to extend the resistive line of the generating unit along the circumferential direction [50], as shown in Figure 4a. The second way is to add an additional distributing unit, as shown in Figure 4b. In both structures, the center is applied with a voltage through an ITO or metal lines. Although there is a voltage drop on the conductive line, it is trivial compared with that on the generating unit. Therefore, the center voltage can be considered to be approximately equal to the externally applied voltage V_1.

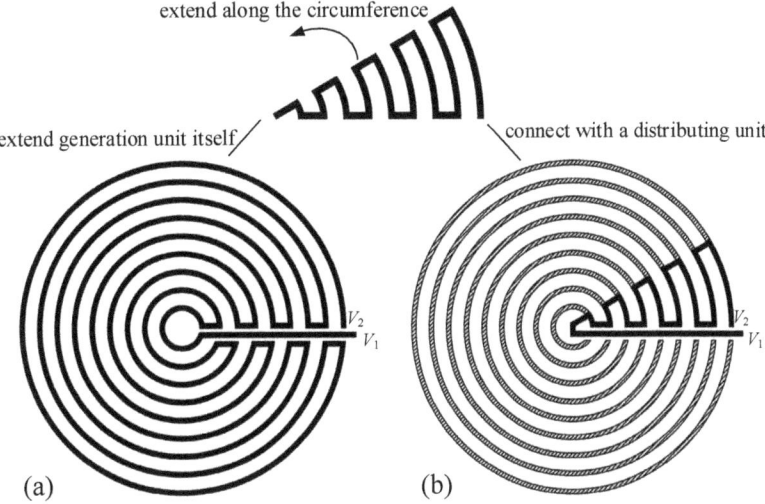

Figure 4. Two kinds of electrode patterns of a circular LC lens extended from the fan-shaped generating unit: (**a**) extend the resistance line of the generating unit itself, or (**b**) connect with a distributing unit.

The above-mentioned structures can produce a parabolic voltage distribution, but when they are used in LC lenses, the voltage distribution is affected by the capacitance effect. To analyze the voltage distribution on the electrode pattern, the LC lens is simplified into an equivalent circuit. The electrode pattern in Figure 4a is taken as an example for the analysis. Each concentric ITO line is a resistor, and they are connected in series. Concentric ITO lines form capacitors with the electrode on another substrate, and they are in parallel with the resistance of the LC material. The circuit of an LC lens is shown in Figure 5.

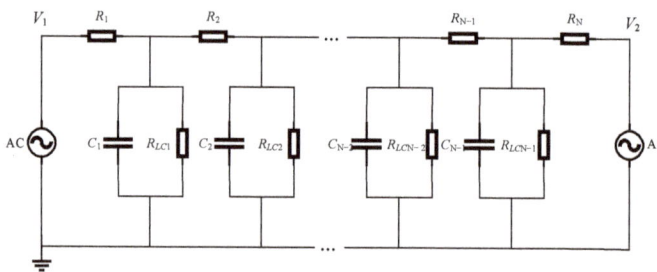

Figure 5. The equivalent circuit of an LC lens based on the electrode pattern in Figure 4a.

The resistance of the concentric ITO line is

$$R_n = \frac{2\pi n(w+d)}{\sigma_e wt} \quad (n = 1, 2, 3, \cdots, N) \tag{2}$$

where σ_e, w, d and t represent the conductivity, width, inter-line gap, and thickness of the ITO electrode, respectively. $N = \lfloor r/(d+w) \rfloor$ is the number of the concentric ITO line, where r is the radius of the lens aperture. The capacitance is

$$C_n = \frac{2\pi n \varepsilon_0 \varepsilon_{LC}(w+d)w}{T} \quad (n = 1, 2, 3, \cdots, N-1) \tag{3}$$

where ε_0 and ε_{LC} represent the vacuum permittivity and the relative permittivity of the LC material, respectively. T is the thickness of the LC layer. The capacitive reactance is expressed as $X_{Cn} = 1/(2\pi f C_n)$, and f is the frequency of the voltage signal. The resistance of the LC material between each concentric ITO line and electrode on another substrate is

$$R_{LCn} = \frac{T}{2\pi n \sigma_{LC}(w+d)w} \quad (n = 1, 2, 3, \cdots, N-1) \tag{4}$$

The resistance of the capacitive reactance in parallel with the resistance of the LC material can be expressed as $R_{Pn} = X_{Cn} R_{LCn}/(X_{Cn} + R_{LCn})$. Then, the circuit in Figure 5 can be further simplified into Figure 6.

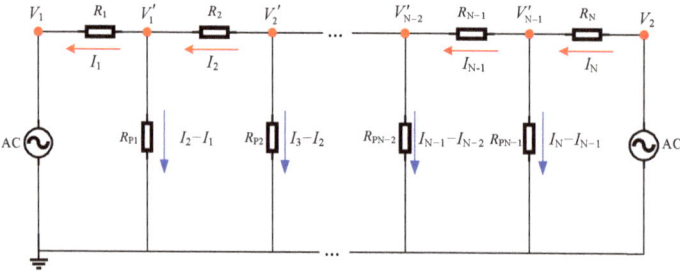

Figure 6. The further simplified circuit from Figure 5.

We assume current magnitude $I_n (n = 1, 2, 3, \cdots, N)$ and current direction. According to Kirchhoff's Voltage Law, the following equations can be obtained

$$\begin{cases} -I_1 R_1 + (I_2 - I_1) R_{P1} - V_1 = 0 \\ -I_N R_N - (I_N - I_{N-1}) R_{PN-1} + V_2 = 0 \\ -I_n R_n + (I_{n+1} - I_n) R_{Pn} - (I_n - I_{n-1}) R_{Pn-1} = 0 \ (n = 2, 3, 4, \cdots, N-1) \end{cases} \quad (5)$$

Solving the equations numerically yields all currents, and the node voltages are expressed as

$$V'_n = (I_{n-1} - I_n) R_{Pn} \quad (n = 1, 2, 3, \cdots, N-1) \quad (6)$$

According to the above analyses, the voltage distribution under any parameters can be simulated, which provides effective guidance for experiments.

3. Experiments and Results

The electrode pattern in Figure 4a is used to fabricate an LC lens, with an aperture size of 2 mm and an LC layer of 30 µm. The parameters of the ITO electrode are $w = d = 5 \mu m$, $t = 20$ nm, and $\sigma_e \approx 10^6$ S/m. The positive nematic LC material used in experiments is HSG28800-100 from HCCH Co. Ltd., for which the conductivity is $\sigma_{LC} \approx 10^{-10}$ S/m, and the optical refractive indices at a temperature of 25 °C and a wavelength of 589 nm are n_e= 1.698, n_o = 1.499 and Δn = 0.199. The dielectric constants at a frequency of 1kHz are $\varepsilon_\perp$ = 3.1, $\varepsilon_{//}$ = 8.0 and $\Delta \varepsilon$ = 4.9. The linear response range of this LC material is measured using the method reported in [47], which is 1.6~2.4 V_{rms}. Then, the driving voltages are controlled in this range to generate a parabolic phase profile. According to Equations (2)–(6), the voltage distributions in the radial direction of the ITO electrode pattern are calculated when V_1 = 1.6 V_{rms}, V_2 = 2.4 V_{rms} and V_1 = 2.4 V_{rms}, V_2 = 1.6 V_{rms}, respectively. The results are shown in Figure 7. The radial voltage distributions of positive and negative lenses basically follow the parabolic profile when the frequency is 1 kHz, while they gradually deviate from the parabolic profile as the frequency increases. Therefore, voltage signals with a frequency of 1 kHz are used to drive the LC lens in experiments.

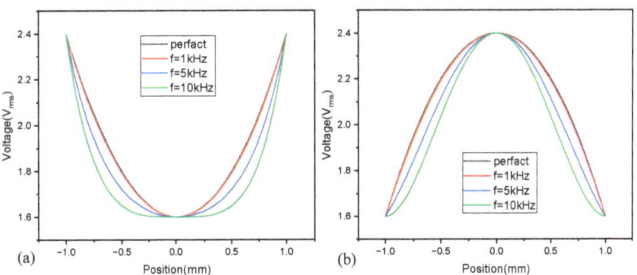

Figure 7. The radial voltage distributions under different frequencies. (**a**) Positive LC lens V_1 = 1.6 V_{rms}, V_2 = 2.4 V_{rms}. (**b**) Negative LC lens V_1 = 2.4 V_{rms}, V_2 = 1.6 V_{rms}.

The electrode pattern shown in Figure 4a is developed on a glass substrate using a single photolithography step. Then, two substrates (one with an electrode pattern and another with a plane ITO electrode) are cleaned with acetone. The polyimide layer is spun and rubbed on substrates to align the nematic director parallel to the substrate surfaces. Then, two substrates are separated by 30 µm spacers and optically aligned facing each other's interior surface in opposing directions. Finally, the LC material is injected into the gap between the two substrates and the LC lens is sealed using the UV curing adhesive.

The fabricated LC lens is inserted into an interferometer to capture the interference fringes, the results of which are shown in Figure 8. The dimensions of each image are 801 × 801 pixel. In Figure 8, the schematic illustration of LC directors corresponding to each lens state is also shown. When $V_1 < V_2$, the tilt angle of the LC director in the border is

Author Contributions: Conceptualization, W.F. and M.Y.; Methodology, W.F. and M.Y.; Software, W.F.; Investigation, W.F., Z.L. and M.Y.; Formal analysis, W.F., Z.L. and M.Y.; Resources, W.F.; Data Curation, W.F., Z.L. and H.L.; Writing—review and editing, W.F. and M.Y.; Supervision, M.Y.; Project administration, M.Y., W.F. and Z.L.; Funding acquisition, M.Y., W.F. and Z.L.; All authors have read and agreed to the published version of the manuscript.

Funding: Sichuan Province Science and Technology Support Program (2021YJ0102).

Institutional Review Board Statement: Not applicable.

Informed Consent Statement: Not applicable.

Data Availability Statement: Not applicable.

Conflicts of Interest: The authors declare no conflict of interest.

References

1. Kumar, A.; Singh, D.P.; Singh, G. Recent progress and future perspectives on carbon-nanomaterial-dispersed liquid crystal composites. *J. Phys. D Appl. Phys.* **2021**, *55*, 083002. [CrossRef]
2. Khoo, I.C. Nonlinear optics of liquid crystalline materials. *Phys. Rep.* **2009**, *471*, 221–267. [CrossRef]
3. Choudhary, A.; Singh, G.; Biradar, A.M. Advances in gold nanoparticle–liquid crystal composites. *Nanoscale* **2014**, *6*, 7743–7756. [CrossRef] [PubMed]
4. Singh, G.; Fisch, M.; Kumar, S. Emissivity and electrooptical properties of semiconducting quantum dots/rods and liquid crystal composites: A review. *Rep. Prog. Phys.* **2016**, *79*, 056502. [CrossRef] [PubMed]
5. Singh, G. Recent advances on cadmium free quantum dots-liquid crystal nanocomposites. *Appl. Mater. Today* **2020**, *21*, 100840. [CrossRef]
6. Sato, S. Liquid-Crystal Lens-Cells with Variable Focal Length. *Jpn. J. Appl. Phys.* **1979**, *18*, 1679–1684. [CrossRef]
7. Ye, M.; Sato, S. Liquid crystal lens with focus movable along and off axis. *Opt. Commun.* **2003**, *225*, 277–280. [CrossRef]
8. Pishnyak, O.; Sato, S.; Lavrentovich, O.D. Electrically tunable lens based on a dual-frequency nematic liquid crystal. *Appl. Opt.* **2006**, *45*, 4576–4582. [CrossRef]
9. Lin, Y.-H.; Chen, H.-S.; Lin, H.-C.; Tsou, Y.-S.; Hsu, H.-K.; Li, W.-Y. Polarizer-free and fast response microlens arrays using polymer-stabilized blue phase liquid crystals. *Appl. Phys. Lett.* **2010**, *96*, 113505. [CrossRef]
10. Lin, H.-C.; Lin, Y.-H. An Electrically Tunable Focusing Pico-Projector Adopting a Liquid Crystal Lens. *Jpn. J. Appl. Phys.* **2010**, *49*, 102502. [CrossRef]
11. Ren, H.; Xu, S.; Wu, S.-T. Polymer-stabilized liquid crystal microlens array with large dynamic range and fast response time. *Opt. Lett.* **2013**, *38*, 3144–3147. [CrossRef] [PubMed]
12. Kawamura, M.; Nakamura, K.; Sato, S. Liquid-crystal micro-lens array with two-divided and tetragonally hole-patterned electrodes. *Opt. Express* **2013**, *21*, 26520–26526. [CrossRef] [PubMed]
13. Algorri, J.F.; Urruchi, V.; García-Cámara, B.; Sánchez-Pena, J.M. Liquid Crystal Lensacons, Logarithmic and Linear Axicons. *Materials* **2014**, *7*, 2593–2604. [CrossRef] [PubMed]
14. Algorri, J.F.; del Pozo, V.U.; Sanchez-Pena, J.M.; Oton, J.M. An Autostereoscopic Device for Mobile Applications Based on a Liquid Crystal Microlens Array and an OLED Display. *J. Disp. Technol.* **2014**, *10*, 713–720. [CrossRef]
15. Chang, Y.-C.; Jen, T.-H.; Ting, C.-H.; Huang, Y.-P. High-resistance liquid-crystal lens array for rotaTable 2D/3D autostereoscopic display. *Opt. Express* **2014**, *22*, 2714–2724. [CrossRef] [PubMed]
16. Shibuya, G.; Yoshida, H.; Ozaki, M. High-speed driving of liquid crystal lens with weakly conductive thin films and voltage booster. *Appl. Opt.* **2015**, *54*, 8145–8151. [CrossRef]
17. Algorri, J.F.; Urruchi, V.; Bennis, N.; Sanchez-Pena, J.M.; Oton, J.M. Cylindrical Liquid Crystal Microlens Array With Rotary Optical Power and Tunable Focal Length. *IEEE Electron Device Lett.* **2015**, *36*, 582–584. [CrossRef]
18. Zhang, Y.; Weng, X.; Liu, P.; Wu, C.; Sun, L.; Yan, Q.; Zhou, X.; Guo, T. Electrically high-resistance liquid crystal micro-lens arrays with high performances for integral imaging 3D display. *Opt. Commun.* **2020**, *462*, 125299. [CrossRef]
19. Ye, M.; Sato, S. Optical Properties of Liquid Crystal Lens of Any Size. *Jpn. J. Appl. Phys.* **2002**, *41*, L571–L573. [CrossRef]
20. Ye, M.; Wang, B.; Sato, S. Driving of Liquid Crystal Lens without Disclination Occurring by Applying In-Plane Electric Field. *Jpn. J. Appl. Phys.* **2003**, *42*, 5086–5089. [CrossRef]
21. Wang, B.; Ye, M.; Sato, S. Liquid crystal lens with focal length variable from negative to positive values. *IEEE Photonics Technol. Lett.* **2005**, *18*, 79–81. [CrossRef]
22. Ye, M.; Wang, B.; Kawamura, M.; Sato, S. Fast switching between negative and positive power of liquid crystal lens. *Electron. Lett.* **2007**, *43*, 474–476. [CrossRef]
23. Lin, H.-C.; Lin, Y.-H. A fast response and large electrically tunable-focusing imaging system based on switching of two modes of a liquid crystal lens. *Appl. Phys. Lett.* **2010**, *97*, 063505. [CrossRef]
24. Lin, H.-C.; Chen, M.-S.; Lin, Y.-H. A Review of Electrically Tunable Focusing Liquid Crystal Lenses. *Trans. Electr. Electron. Mater.* **2011**, *12*, 234–240. [CrossRef]

25. Lin, Y.-H.; Wang, Y.-J.; Reshetnyak, V. Liquid crystal lenses with tunable focal length. *Liq. Cryst. Rev.* **2017**, *5*, 111–143. [CrossRef]
26. Naumov, A.F.; Loktev, M.Y.; Guralnik, I.R.; Vdovin, G. Liquid-crystal adaptive lenses with modal control. *Opt. Lett.* **1998**, *23*, 992–994. [CrossRef]
27. Naumov, A.F.; Love, G.D.; Loktev, M.Y.; Vladimirov, F.L. Control optimization of spherical modal liquid crystal lenses. *Opt. Express* **1999**, *4*, 344–352. [CrossRef]
28. Ye, M.; Wang, B.; Sato, S. Realization of liquid crystal lens of large aperture and low driving voltages using thin layer of weakly conductive material. *Opt. Express* **2008**, *16*, 4302–4308. [CrossRef]
29. Kotova, S.P.; Patlan, V.V.; Samagin, S.A. Tunable liquid-crystal focusing device. 1. Theory. *Quantum Electron.* **2011**, *41*, 58–64. [CrossRef]
30. Kotova, S.P.; Patlan, V.V.; Samagin, S.A. Tunable liquid-crystal focusing device. 2. Experiment. *Quantum Electron.* **2011**, *41*, 65–70. [CrossRef]
31. Algorri, J.F.; Morawiak, P.; Zografopoulos, D.C.; Bennis, N.; Spadlo, A.; Rodriguez-Cobo, L.; Jaroszewicz, L.R.; Sanchez-Pena, J.M.; Lopez-Higuera, J.M. Multifunctional light beam control device by stimuli-responsive liquid crystal micro-grating structures. *Sci. Rep.* **2020**, *10*, 13806. [CrossRef] [PubMed]
32. Algorri, J.F.; Morawiak, P.; Bennis, N.; Zografopoulos, D.C.; Urruchi, V.; Rodriguez-Cobo, L.; Jaroszewicz, L.R.; Sanchez-Pena, J.M.; Lopez-Higuera, J.M. Positive-negative tunable liquid crystal lenses based on a microstructured transmission line. *Sci. Rep.* **2020**, *10*, 10153. [CrossRef] [PubMed]
33. Algorri, J.F.; Morawiak, P.; Zografopoulos, D.C.; Bennis, N.; Spadlo, A.; Rodriguez-Cobo, L.; Jaroszewicz, L.R.; Sanchez-Pena, J.M.; Lopez-Higuera, J.M. Cylindrical and Powell Liquid Crystal Lenses With Positive-Negative Optical Power. *IEEE Photonics Technol. Lett.* **2020**, *32*, 1057–1060. [CrossRef]
34. Bennis, N.; Jankowski, T.; Morawiak, P.; Spadlo, A.; Zografopoulos, D.C.; Sánchez-Pena, J.M.; López-Higuera, J.M.; Algorri, J.F. Aspherical liquid crystal lenses based on a variable transmission electrode. *Opt. Express* **2022**, *30*, 12237–12247. [CrossRef] [PubMed]
35. Algorri, J.; Zografopoulos, D.; Rodríguez-Cobo, L.; Sánchez-Pena, J.; López-Higuera, J. Engineering Aspheric Liquid Crystal Lenses by Using the Transmission Electrode Technique. *Crystals* **2020**, *10*, 835. [CrossRef]
36. Pusenkova, A.; Sova, O.; Galstian, T. Electrically variable liquid crystal lens with spiral electrode. *Opt. Commun.* **2021**, *508*, 127783. [CrossRef]
37. Stevens, J.; Galstian, T. Electrically tunable liquid crystal lens with a serpentine electrode design. *Opt. Lett.* **2022**, *47*, 910–912. [CrossRef]
38. Wang, B.; Ye, M.; Honma, M.; Nose, T.; Sato, S. Liquid Crystal Lens with Spherical Electrode. *Jpn. J. Appl. Phys.* **2002**, *41*, L1232–L1233. [CrossRef]
39. Lin, C.-H.; Chen, C.-H.; Chiang, R.-H.; Jiang, I.-M.; Kuo, C.-T.; Huang, C.-Y. Dual-Frequency Liquid-Crystal Lenses Based on a Surface-Relief Dielectric Structure on an Electrode. *IEEE Photonics Technol. Lett.* **2011**, *23*, 1875–1877. [CrossRef]
40. Wang, B.; Ye, M.; Sato, S. Lens of electrically controllable focal length made by a glass lens and liquid-crystal layers. *Appl. Opt.* **2004**, *43*, 3420–3425. [CrossRef]
41. Ren, H.; Fan, Y.-H.; Gauza, S.; Wu, S.-T. Tunable-focus flat liquid crystal spherical lens. *Appl. Phys. Lett.* **2004**, *84*, 4789–4791. [CrossRef]
42. Ren, H.; Wu, S.-T. Adaptive liquid crystal lens with large focal length tunability. *Opt. Express* **2006**, *14*, 11292–11298. [CrossRef] [PubMed]
43. Huang, Y.-P.; Chen, C.-W.; Shen, T.-C. High resolution autostereoscopic 3D display with scanning multi-electrode driving liquid crystal (MeD-LC) Lens. In *2009 SID International Symposium Digest of Technical Papers, Vol XL, Books I–III*; Society for Information Display: San Jose, CA, USA, 2009; pp. 336–339.
44. Beeckman, J.; Yang, T.-H.; Nys, I.; George, J.P.; Lin, T.-H.; Neyts, K. Multi-electrode tunable liquid crystal lenses with one lithography step. *Opt. Lett.* **2018**, *43*, 271–274. [CrossRef] [PubMed]
45. Shen, T.; Zheng, J.; Shen, C.; Wang, X.; Liu, Y. One-dimensional straight-stripe-electrode tunable self-focusing cylindrical liquid crystal lens to realize the achromatic function in time domain. *Optik* **2022**, *260*, 168962. [CrossRef]
46. Li, L.; Bryant, D.; Bos, P.J. Liquid crystal lens with concentric electrodes and inter-electrode resistors. *Liq. Cryst. Rev.* **2014**, *2*, 130–154. [CrossRef]
47. Feng, W.; Liu, Z.; Ye, M. Positive-negative tunable cylindrical liquid crystal lenses. *Optik* **2022**, *266*, 169613. [CrossRef]
48. Feng, W.; Ye, M. Positive-Negative Tunable Liquid Crystal Lens of Rectangular Aperture. *IEEE Photonics Technol. Lett.* **2022**, *34*, 795–798. [CrossRef]
49. Feng, W.; Liu, Z.; Ye, M. Liquid crystal lens array with positive and negative focal lengths. *Opt. Express* **2022**, *30*, 28941. [CrossRef]
50. Feng, W.; Liu, Z.Q.; Xu, L.; Li, H.; Ye, M. A design method of high-performance liquid crystal lens. *Acta Opt. Sin.* **2023**, *43*, 0223001.

Disclaimer/Publisher's Note: The statements, opinions and data contained in all publications are solely those of the individual author(s) and contributor(s) and not of MDPI and/or the editor(s). MDPI and/or the editor(s) disclaim responsibility for any injury to people or property resulting from any ideas, methods, instructions or products referred to in the content.

Article

End-of-Life Liquid Crystal Displays Recycling: Physico-Chemical Properties of Recovered Liquid Crystals

Idriss Moundoungou [1], Zohra Bouberka [1,2], Guy-Joël Fossi Tabieguia [1], Ana Barrera [1,*], Yazid Derouiche [1,3], Frédéric Dubois [4], Philippe Supiot [1], Corinne Foissac [1] and Ulrich Maschke [1]

[1] Unité Matériaux et Transformations (UMET), UMR 8207, CNRS, INRAE, Centrale Lille, Université de Lille, 59000 Lille, France
[2] Laboratoire Physico-Chimie des Matériaux-Catalyse et Environnement (LPCM-CE), Université des Sciences et de la Technologie d'Oran Mohamed Boudiaf (USTOMB), BP 1505, El M'naouer, Oran 31000, Algeria
[3] Laboratoire Physico-Chimie des Matériaux et Environnement (LPCME), Université ZIANE Achour, Djelfa 17000, Algeria
[4] UR 4476, Unité de Dynamique et Structure des Matériaux Moléculaires (UDSMM), Université du Littoral Côte d'Opale, 59379 Dunkerque, France
* Correspondence: ana-luisa.barrera-almeida@univ-lille.fr; Tel.: +33-3-20-33-63-81

Citation: Moundoungou, I.; Bouberka, Z.; Fossi Tabieguia, G.-J.; Barrera, A.; Derouiche, Y.; Dubois, F.; Supiot, P.; Foissac, C.; Maschke, U. End-of-Life Liquid Crystal Displays Recycling: Physico-Chemical Properties of Recovered Liquid Crystals. *Crystals* 2022, 12, 1672. https://doi.org/10.3390/cryst12111672

Academic Editors: Zhenghong He and Yuriy Garbovskiy

Received: 2 November 2022
Accepted: 17 November 2022
Published: 19 November 2022

Publisher's Note: MDPI stays neutral with regard to jurisdictional claims in published maps and institutional affiliations.

Copyright: © 2022 by the authors. Licensee MDPI, Basel, Switzerland. This article is an open access article distributed under the terms and conditions of the Creative Commons Attribution (CC BY) license (https://creativecommons.org/licenses/by/4.0/).

Abstract: This report focuses particularly on liquid crystals display (LCD) panels because they represent a significant amount of all WEEE collected. Technologies involving liquid crystals (LCs) have enjoyed considerable success since the 1970s in all fields of LC displays (LCDs). This currently provokes the problem of waste generated by such equipment. Based on current statistical data, the LC amount represents approximately 1.3 g for a 35-inch diameter LCD panel unit possessing a total weight of 15 kg. In France, a recent study revealed LCD waste to represent an average of 5.6 panels per household. This represents an important quantity of LCs, which are generally destroyed by incineration or washed out with detergents during the recycling processes of end-of-life (EOL) LCDs. Hence, the aim of this study is to show that it is possible to remove LC molecules from EOL-LCD panels with the goal of valorizing them in new sectors. EOL-LCD panels have undergone various stages of dismantling, chemical treatments and characterization. The first stage of manual dismantling enables the elimination of the remaining physical components of the panels to process LC molecules only, sandwiched between the two glass plates. Mechanical treatment by scraping allows us to obtain a concentrate of LCs. The results obtained from chemical and physical techniques show that these molecules retain the characteristics essential for their operation in the field of optical and electro-optical devices. As the use of LCD surfaces continues to rise significantly, the amounts and economic stakes are huge, fully justifying the development of an LC recovery process for used panels. Many potential uses have been identified for these LC molecules: in new flat LCD panels after purification of the LCs concentrate, in PDLC systems, as lubricants or in thermal applications.

Keywords: WEEE; liquid crystal displays; liquid crystals; recycling

1. Introduction

In 2003, the European Waste Electrical and Electronic Equipment (WEEE) Directive (2002/96/EC) was created, with the main objective of reducing the production of WEEE and requiring the disassembly of all liquid crystal displays (LCDs) with an area over 100 cm^2 [1]. Because EEE and WEEE may contain hazardous chemicals, they are subject to the provisions of the European RoHS Directive (2002/95/EC) (restriction on the use of certain hazardous substances) [2], which aims to reduce the number of hazardous materials in EEE. In France, these European directives (WEEE and RoHS) were transposed by decree and codified in the Environmental Code in 2005 (2005-829) [3]. Then, they were amended in 2012 (2012-617) and 2014 (2014-928) [4,5]. There are five types of WEEE treatment classified below by order of priority defined by the regulations: preparation for reuse:

reuse of the whole equipment; reuse of parts: reuse of parts or sub-assemblies of the equipment; material recycling: recycling of the material; energy recovery: incineration with energy recovery; disposal: disposal without recovery (landfill, incineration without energy recovery) [6].

Technologies involving liquid crystals (LCs) have enjoyed considerable success since the 1970s in all fields of LCDs. However, the success of technologies involving LCs generates large quantities of LCD waste representing a significant quantity of LC mixtures abandoned in end-of-life (EOL)-LCD panels. A study conducted by the "Conseil Supérieur de l'Audiovisuel" (CSA) showed that in 2018, an average waste of 5.6 panels was determined per French household, which is constantly increasing [7].

LCs are organic molecules that exhibit intermediate states between the crystalline solid state characterized by a long-range order and the isotropic liquid, where there is no order [8]. These states are called mesophases [9]. LCs used in LCD panels form a mixture of molecules of polycyclic aromatic hydrocarbons and alkyl, alkoxy, chloro, fluoro or carboxyl chains [10]. Many practical considerations must be met by an LC mixture prior to its use in a display device. The essential behavior of these blends is their sensitivity to an applied electric field. Fluorinated groups bonded to aromatic rings can increase their sensitivity to electric fields [11], i.e., LC molecules switch in response to an electric field in the direction of the passage of light. Conversely, when the electric field is removed, LC molecules are likely to return to their original position due to inherent elastic forces.

LC mixtures destinated for display devices should provide a wide operating temperature range, typically between $-30\ °C$ and $+80\ °C$, allowing the device to operate in both summer and winter climatic conditions all over the world. The LC mixture should have high chemical, electrochemical and physicochemical stability. It should be stable in air, moisture and against oxidative conditions. These last properties are achieved by adding to the mixture other compounds such as antioxidants, stabilizers and UV stabilizers. Finally, the LC mixtures must have desired refractive indices, dielectric constants, elastic constants and a low viscosity, which allows seeing fast-moving images. For the latter property, a family of phenylcyclohexanes can be chosen with linked alkoxyl groups.

As shown in Figure 1, a LCD module consists of multiple layers. Considered as a one-dimensional light source, lamps are often placed linearly on the bottom chassis with reflective sheets. A diffuser plate and a diffuser sheet are mounted over the lamps to generate a uniform distribution of light so that backlights can provide a two-dimensional distribution of light. In addition to spreading light from the lamps, the diffuser plate made of polymethylmethacrylate (PMMA) also functions as a support for holding up diffuser sheets. A prism sheet is used to increase the brightness measured normally to the surface after the light is scattered through the diffuser sheet. Since the light passing through the prism sheet is not polarized, a reflective polarizer is added to recycle light lost to absorption by controlling polarization and reflection [12].

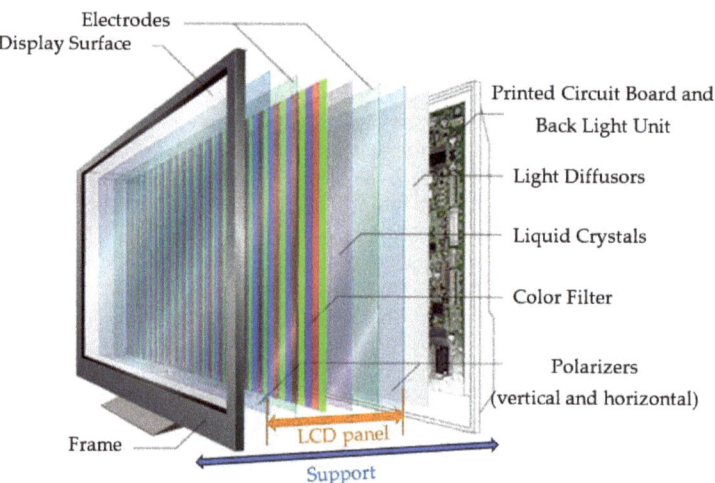

Figure 1. Structure of an LCD-TV module (from [13]).

The structure of the LCD central module comprising LCs is shown schematically in Figure 2. LC mixtures are sandwiched between two parallel plates (usually glass), exhibiting a layer thickness varying between 3 and 5 µm. The glass plates are coated on one side by a thin layer of ITO (Indium Tin Oxide) which represents a thickness between 0.1 and 0.3 µm [14]. Indeed, to extract only the LCD panel containing LCs from the whole TV apparatus, various components must be successfully removed, generally by manual operations. For instance, the plastic TV cover shell must be removed by unscrewing the front and rear parts. It should be remembered that there are hundreds of LC compounds currently employed in LCDs, and a typical LCD panel could contain as many as 25 different compounds that are mixed together to form a white, opaque liquid that flows easily. Each of these LC compounds possesses different physical and optical characteristics [15,16].

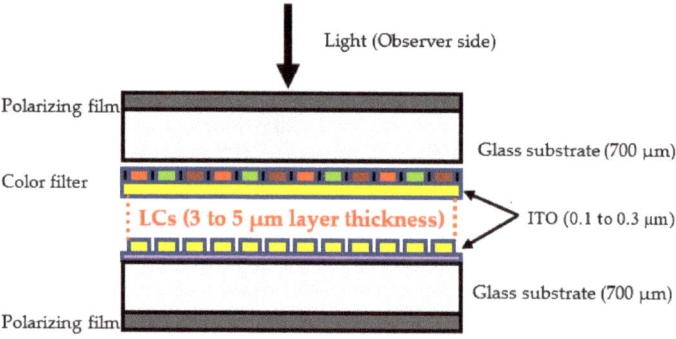

Figure 2. Schematic localization of LCs in the LCD panels.

The use of LCD panels in plate panels of TV or monitors currently provokes the problem of waste treatments generated by such material. Processing technologies, including the recovery of LCs, are rare or under development [17–19]. One of the main procedures of the processing of waste from LCD panels includes manual dismantling of the modules, during which the backlight tubes containing mercury are separated and directed towards specific pathways for processing tubes and discharge lamps [20–23]. Following this procedure, LCs are generally rejected or destroyed by incineration or washing detergents [24]. This seems to be justified by the fact that other materials such as plastics, metals and electronic components account for more than 90% by weight of all components in a typical LCD panel, while LCs represent less than 0.1%. The LC amount represents approximately 1.3 g for an LCD panel unit with a 35-inch diameter, possessing a total weight of 15 kg [25]. This quantity must be reviewed in the context of a real dismantling project to better understand the economic potential of recycling used LCs. Indeed, if we consider that a large number of EOL-LCD panels will accumulate in waste disposal centers, the overall amount of LC is no longer negligible. Thus, the aim of this work is to investigate the possibility of removing LC molecules from EOL-LCD panels by a simple method and then check for recovery of their physico-chemical and optical properties.

It should be mentioned that economic issues play an important role in developing the industrial recovery of LCs. An adapted product specification must be established on the basis of extended academic research, aiming toward the valorization of recycled LC mixtures.

2. Materials and Methods

2.1. Materials: EOL-LCD Panels

The EOL-LCD panels that have been permitted to achieve this study were obtained from the ENVIE^{2}E Company based in the north of France. The business fields of ENVIE^{2}E are recovering, sorting, reusing, recycling and upgrading WEEE and other materials. This paper is based on the treatment of 35 EOL-LCD panels older than five years. These were dismantled manually to recycle LC molecules. Industrial LCD recycling facilities, such as ENVIE^{2}E, must ensure the safety of workers. One of the most important issues during dismantling is the presence of mercury in the old tubes used as light sources. The dismantling process is carried out in a controlled atmosphere (fume cupboards) coupled with a filtration system, allowing the capture of volatile species using activated carbon. Authorized companies carry out the regeneration of the activated carbon in order to analyze the captured species that may contribute to environmental pollution. In addition, the operators wear personal protective equipment adapted to their work (safety shoes and glasses, helmet, gloves, etc.).

2.2. Extraction of LCs from EOL-LCD Panels

In order to extract only the LCD panel containing LCs from the whole TV apparatus, various components were successively removed following a specific order by manual operations (Figure 3). The plastic TV cover shell was removed by unscrewing the front and rear parts. During disassembly, there is no danger of damaging the backlight tubes incorporating mercury since they are fixed at the rear plastic plates. The LCs sandwiched between the two ITO-coated glass plates could thus be removed before reaching backlight tubes. The processing of these tubes and polymers and the recovery of other parts will not be discussed in this report.

The sandwiched glass plates were first manually opened using a scalpel or spatula (Figure 3a) to separate them (Figure 3b). This step allows access to the LC mixtures (Figure 3c,d), which show translucent and slightly viscous aspects. To extract all LCs from the glass supports, a plastic squeegee was used on the whole glass surface of the plates to scrape LCs molecules (Figure 3c). This operation was repeated several times to ensure optimal extraction after a visual check (Figure 3d). The thickness of each LCD screen, comprised of the two glass substrates and the LC layer, was measured with a mechanical micrometer. Thicknesses of the LC layer were deduced from the subtraction of the thickness of the glass plates from the total value of the LCD screen, yielding values between 3 and 5 µm.

Thus, the obtained LCs were ready to be analyzed by several physico-chemical and optical techniques.

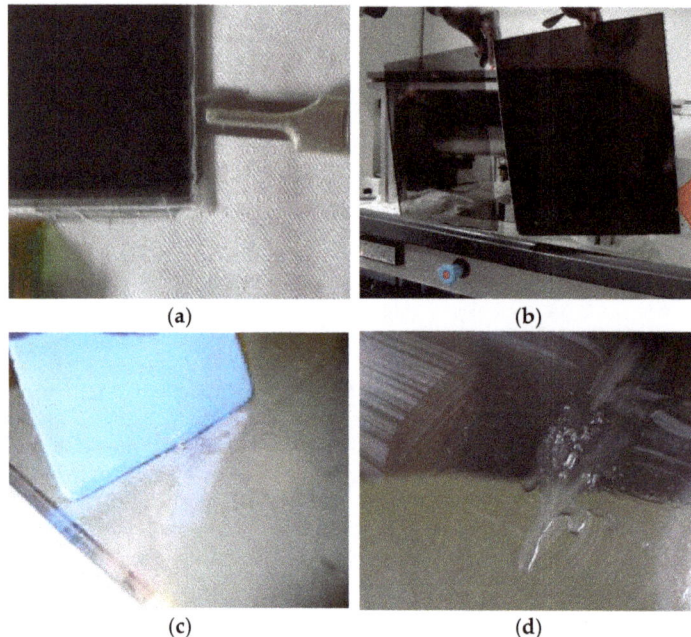

Figure 3. The opening stages of an LCD panel giving access to LCs. (**a**,**b**) Manual opening of the sandwiched glass plates using a scalpel. (**c**) Separation of the two glass plates, on the left-hand side, front of the panel consisting of color filters and black matrix; on the right-hand side, back side. (**d**) Collected LCs molecules with translucent and slightly viscous aspects.

2.3. Characterization Techniques

To ensure that the physico-chemical and optical characteristics of LCs are maintained despite the age and the treatment applied on EOL-LCD panels, characterization studies were carried out on the extracted products. This report focuses on the results of four individual EOL-LCD panels, but several characterizations realized on other samples provided the same conclusion [26]: A (diagonal size 31 inches), B (27 inches), C (26 inches) and D (32 inches). Each capital letter (A to D) represents a different manufacturer. All samples were treated in the same manner to proceed to the extraction of LC mixtures.

2.3.1. Determination of Molecular Structures by FTIR, ^{1}H NMR and GC-MS

Chemical structures of the LCs molecules were obtained using Fourier transform infrared (FTIR), proton nuclear magnetic resonance (^{1}H NMR) spectroscopies, and gas chromatography coupled with mass spectroscopy (GC-MS). FTIR spectra were recorded in the transmission mode using a Perkin Elmer Frontier model. The number of accumulated scans was 16, with a spectral resolution of 4 cm^{-1}. ^{1}H NMR spectra were recorded using an FT-NMR (300 MHz) Bruker instrument with Tetramethylsilane as the internal standard at room temperature in CDCl$_3$. The recovered LC mixtures were dried at 70 °C in a vacuum oven and analyzed by GC-MS. The GC chromatograms and their associated mass spectra were obtained using a Perkin Elmer Clarus 680 gas chromatograph coupled with a Clarus 600T mass detector (PerkinElmer, Shelton, CT, USA). A fused silica capillary column (Elite-5 (5% Diphenyl) Dimethylpolysiloxane, 30 m × 0.53 mm (internal diameter), film thickness 0.5 µm; temperature of the column 70 °C) was employed. The mass spectrometer was equipped with an electronic ionization (EI) source and a quadripole mass analyzer (QSM). The injector temperature was 300 °C, and a 10 µL syringe was used for injections of 0.2 µL. Helium was the carrier gas, applying a constant flow of 1.5 mL/min. The ion source and interface temperatures were maintained at 180 and 300 °C, respectively. The furnace temperature was held for 15 min at 70 °C, then raised to 150 °C with a rate of 20 °C/min followed by an isotherm of 10 min, then increased to 240 °C followed by an isotherm of 10 min. The retention times (t_R) were determined for each separated molecule.

2.3.2. Determination of Thermal and Optical Properties

Thermal and optical properties of LC mixtures extracted from EOL-LCD panels were examined by differential scanning calorimetry (DSC), polarized optical microscopy (POM) and thermo gravimetrical analysis (TGA). DSC experiments were carried out using a Pyris Diamond DSC calorimeter at a heating rate of 10 °C/min under continuous nitrogen flow. The sample weight was approximately 7 mg, and the data were evaluated from the second heating ramps. The thermo-microscopy studies were performed on a POM Olympus BX41 equipped with a digital camera conjugated with a PC. The system comprises a Linkam heating/cooling stage LTS 350 together with a temperature-controlling unit TMS 94. The temperature was increased in a stepwise manner, the typical size of one step being generally 2 °C. Thermogravimetric analysis was conducted on a TA Instruments Q5000 based on a heating rate of 10 °C/min from T = 20 °C to T = 600 °C, under continuous air flow with a sample weight of about 15 mg.

3. Results and Discussions

Since LCD screens contain various amounts of different unknown LC molecules (and others), and these LC mixtures were developed as a function of specific applications, it was not appropriate to investigate original commercial LC mixtures in order to compare their physico-chemical properties with those from the collected EOL-LC mixtures.

3.1. Mass Balance of Extracted LCs

The results from the mass balance of LCs recovered by scraping are presented in Figure 4. High dispersion of data was observed, and limited amounts of less than 1 g per panel for two reasons. The first concerns the status of approved monitors on the dismantling site. Many panels were found in a state of mechanical damage (crack, breakage of the glass panel, LC exposed to atmosphere). This degradation involves lower LC extracting from surfaces than expected. In addition, as the LCs are in the form of flowable gel, logistic and handling operations could cause them to be lost. Secondly, the efficiency of recovery depends on how the surface of the glass plate of the LCD panel was scratched: more or less strong in an area or several times in the same area. Thus, transport operations for EOL-LCDs must take into account their fragility. Similarly, recovery techniques must be implemented with an efficient process. This last point has been investigated in this report [27]. .

To conduct a detailed characterization of LCs from these EOL-LCD panels, a selection of four undamaged panels with good general status was conducted (Table 1). The results of this selection show that the LCs mass increases with the size of the panel. Not surprisingly, the largest mass was obtained with panel C having a diagonal of 81.2 cm (32 inches), and the lowest amount corresponded to panel D with a diagonal of 58.4 cm (23 inches). Secondly, in theory, a panel exhibiting a diagonal of 32 inches corresponds to a total amount of 1.2 g of LCs if one considers an LC layer thickness corresponding to 5 µm [12]. It appears that the quantity of extracted LCs was lower than the theoretical one but better than in the case of damaged panels. Thus, the extraction yield reached approximately 70–80% of the theoretical value.

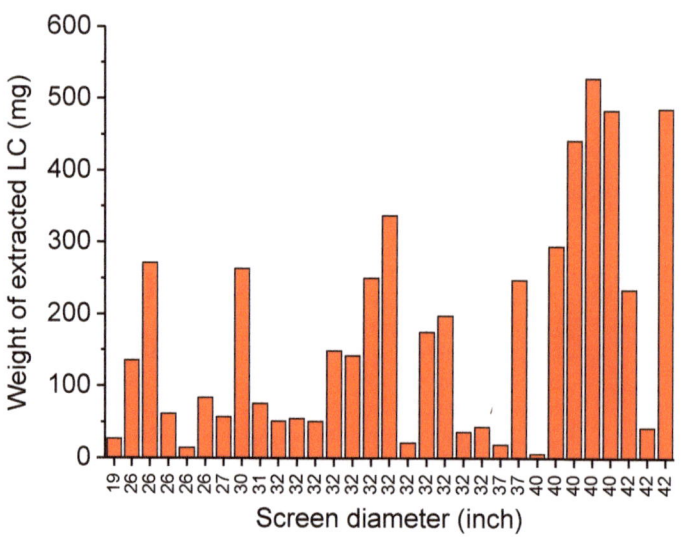

Figure 4. Weight of LCs mixtures extracted manually from 31 EOL-LCD panels.

Table 1. Amount of LCs mixtures extracted from selected EOL-LCD panels A, B, C and D.

	D-LCs	B-LCs	A-LCs	C-LCs
Diameter (inch) of the LCD panel	23	26	32	32
Mass (mg) of recovered LC	347.6	575.4	754.8	873.5

3.2. Chemical Characterization of LCs Mixtures

FTIR spectroscopy provides detailed information regarding the assignment distributions of specific functional groups of LCs extracted from EOL-LCD panels (Figure 5).

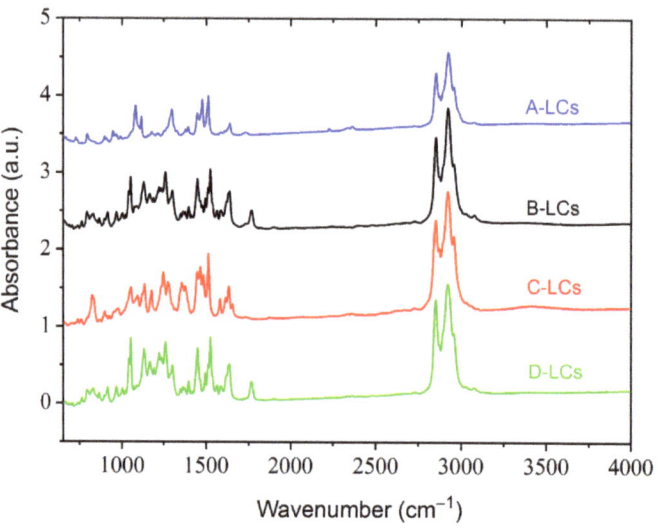

Figure 5. FTIR results from LC mixtures of LCDs A, B, C and D.

3.2.1. Identification of Aromatic Groups and Possible Substitutions

For all LCs molecules stretching vibration bands characteristic of aromatic C–H (small peaks of around 3020 and 3060 cm^{-1}) and C=C (peaks at 1470, 1510 and 1625 cm^{-1}), bonds were detected. Detailed investigation of the 700–800 cm^{-1} band also provided some information regarding the substitutions on the aromatic cores. In the case of A-LCs, the existence of some bands at 720 and 793 cm^{-1} indicated that the aromatic cores were, respectively, mono- and para-disubstituted. This was also the case for the D-LCs illustrated by the vibration at 759 cm^{-1} (mono-substituted) and 789–862 cm^{-1} (para-disubstituted). The B-LCs contained ortho- (758 cm^{-1}) and para- (790–850 cm^{-1}) disubstituted aromatic products. Finally, C-LCs were characterized by some bands at 734 cm^{-1} (mono-substituted), 742 cm^{-1} (ortho-disubstituted) and a pronounced peak at 821 cm^{-1} (para-disubstituted).

3.2.2. Alkyl and Alkane Groups

The presence of methyl groups was illustrated by some bands around 1255–1295 cm^{-1} (for all LCs), 1390 cm^{-1} (A- and B-LCs) and 2943 cm^{-1} (for all LCs). The existence of methylene groups was indicated by some vibration bands at 2850 and 2916 cm^{-1} (for all LCs). Alkane chains (RCH_2CH_3) were characterized by stretching vibrations around 1370–1390 cm^{-1} and a band at 1450 cm^{-1}.

3.2.3. Aliphatic Group Associated with Specific Chemical Bonds

The products were characterized by strong bands at 1050 cm^{-1} (B-LCs, C-LCs and D-LCs) and 1070 cm^{-1} (A-LCs), which correspond to the ether function (R-O-R). Some bands seem to indicate the presence of fluorinated bonds. These appear at 1120–1160 cm^{-1} (C-F) and 1170–1200 cm^{-1} (C-F_3). Additionally, B-LC-, C-LC- and D-LC-mixtures were characterized by a peak at 1580 cm^{-1}, indicating the presence of an N-H bond. The peak at 1760 cm^{-1} was easily distinguishable in the case of B-LCs and D-LCs products but weakly pronounced for A-LCs. This band is characteristic of a C=O bond stretching vibration, which indicates the presence of ester groups. This C=O bond is usually associated with another oxygen to build an ester of the formula R-COO-R'. In this structure, the R' part is often susceptible to carry the dipole moment of the LC molecule. Finally, the A-LC products showed a weak peak at 2221 cm^{-1}, probably corresponding to the presence of a C≡N bond.

The FTIR results corroborated ^{1}H-NMR (Figure 6) and GC-MS (Figure 7) data. Indeed, despite the complexity and the high number of individual molecules in the mixtures used for LCD applications [28], qualitative analysis of spectra allowed us to identify the principal chemical structures of the extracted LCs. Common chemical groups were found for all products. Aromatic (Ar) rings corresponded to chemical shifts in the region of 7.7–6.6 ppm. The other common groups were "R-CH_2-O-Ar" (4.2 ppm), "CH_3-O-Ar" (3.8 ppm), "CH_2-Ar" (2.2 ppm) and aliphatic groups (3–0.7 ppm). However, the "-HC=CH-" group (5.8, 4.8 ppm), observed for A-, B- and D-LCs, was absent in the C-LC sample.

The results of FTIR and ^{1}H NMR analysis are summarized in Table 2. Many LCs generally present several common characteristics. LCs are polarizable molecules consisting of a rigid core unit, flexible ends and polar groups. The flexible alkyl chains reduce the melting point, and the mesogenic rigid cores and polar groups provide the anisotropy necessary for the formation of LC phases. Almost all LC molecules contain two or more phenyl rings linked by –COO, –C=C– and other groups, with various flexible terminal groups and a polar end-group (–CN, –NC and F) [28]. Indeed, the samples consist of chemical groups which correspond to a rigid core (aromatic rings), flexible chains (alkyl, alkane, methylene) or polar groups (presence of O, N, F). The presence of different polar groups in these products extracted from EOL-LCD panels responds to the fact that the design of LC molecules must take into account the relative dipole moment and the position of polar groups within the molecule, the overall molecular polarizability and the presence of any stereogenic centers [29].

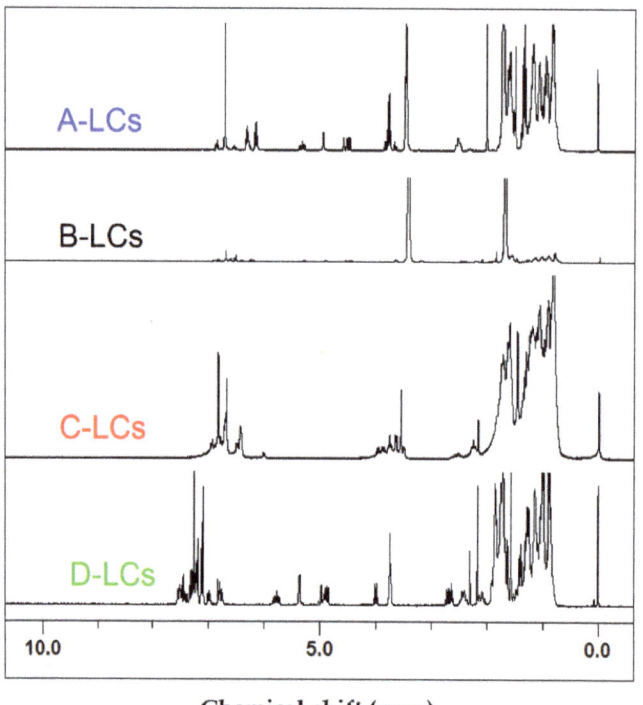

Figure 6. ^{1}H-NMR results from LC mixtures of EOL-LCD A, B, C and D.

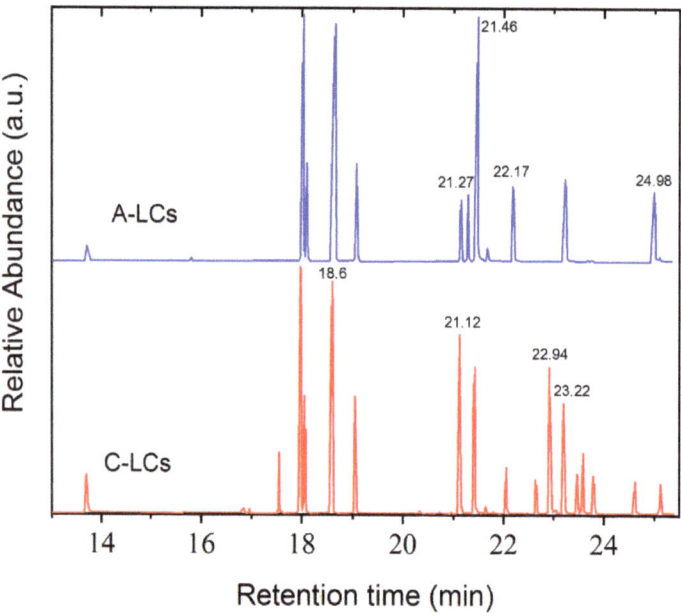

Figure 7. GC results from LC mixtures of EOL-LCD A and C.

Table 2. Correspondence between general LC characteristics and results from FTIR and ^{1}H NMR analysis of EOL-LCDs A, B, C and D.

LC Characteristics	A	B	C	D
Rigid core	Aromatic rings Mono-, para-substituted	Aromatic rings Para-, ortho-substituted	Aromatic rings Mono-, para-, ortho-substituted	Aromatic rings Mono-, para-substituted
Flexible chains	CH3-... ...-CH2-... R...CH2...CH3 -HC=CH-	CH3-... ...-CH2-... R...CH2...CH3 -HC=CH-	CH3-... ...-CH2-... R...CH2...CH3	CH3-... ...-CH2-... R...CH2...CH3 -HC=CH-
Polar groups	C≡N, C-O, C-F/C-F3, C=O	C-O, C-F/C-F3, C=O	C-O, C-F/C-F3,	C-O, C-F/C-F3, C=O

Figure 7 shows the GC chromatograms obtained from two representative LC mixtures, A and C. Each peak corresponds to the presence of molecules in the analyzed samples. Applying the mass spectroscopy database, these peaks could be identified. Table 3 gathers some of the main LC components found in the studied mixtures. The chemical structures of these molecules show mainly aromatic rings, polar groups as well as alkyl chains, thus corroborating the typical structure of nematic LCs possessing permanent dipole moments. The presence of fluorine was detected in these LC molecules, as already observed by FTIR and NMR analysis. Other molecules that do not present a typical LC structure were also identified, i.e., 1-Chloro-4-(4-methyl-4-pentenyl) benzene (t_R = 18 min), 4-Biphenylol diphenyl phosphate (t_R = 19.05 min) and Triphenyl phosphate (t_R = 23.9 min).

Table 3. Assignment of some peaks corresponding to LC molecules of GC chromatograms obtained from LC mixtures A and C.

Retention Time (min)	Assignment
	A-LCs
21.27	4-ethyl-4'-(4-(trifluoromethoxy) phenyl)-1,1'-bi(cyclohexane)
21.46	4-butoxy-2,3-dicyanophenyl 4-((1r,4r)-4-ethylcyclohexyl) benzoate
22.17	4'-propyl-bicyclohexyl-4-carboxylic acid 4-fluoro-phenyl ester

Table 3. *Cont.*

Retention Time (min)	Assignment
24.98	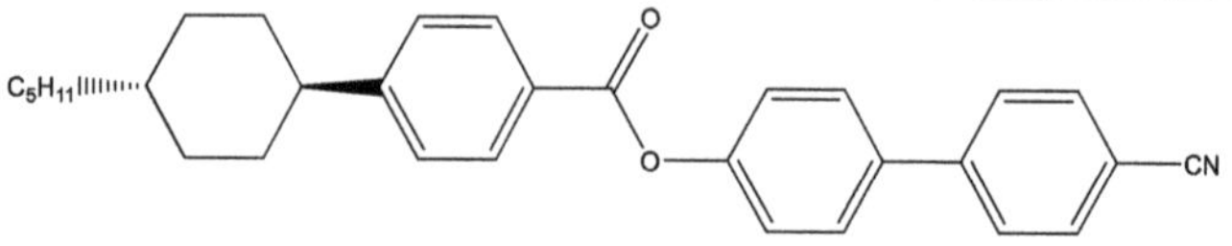 4′-cyano-(1,1′-biphenyl)-4-yl 4-(-4-pentylcyclohexyl) benzoate
	C-LCs
18.60	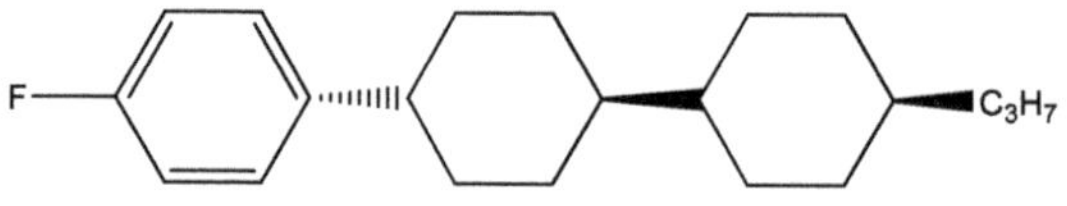 4-(4-Fluoro-phenyl)-4′-propyl-bicyclohexyl
21.12	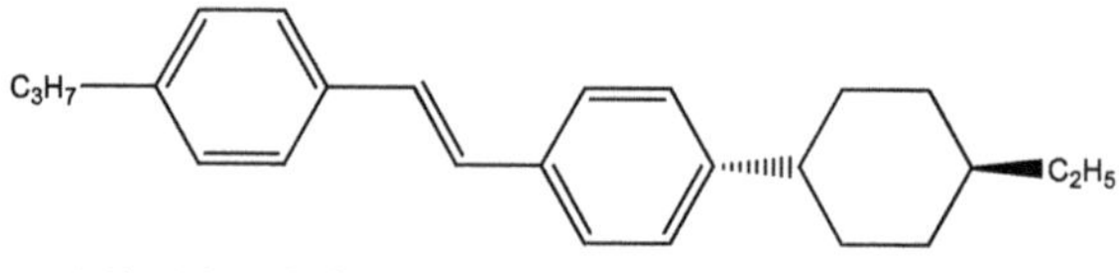 1-((1r,4r)-4-ethylcyclohexyl)-4-((E)-4-propylstyryl) benzene
22.94	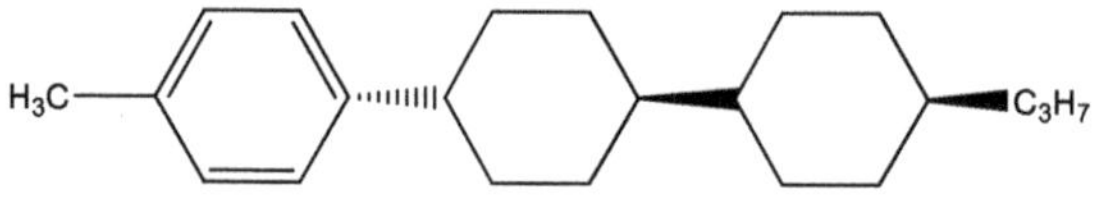 (1s, 1r′, 4R, 4′R)-4-propyl-4′-(p-tolyl)-1,1′-bi(cyclohexane)
23.22	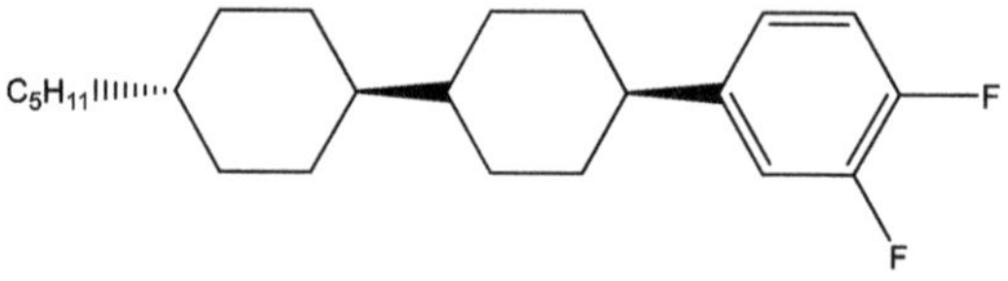 4-(3,4-difluoro-phenyl)-4′-pentyl-1,1′-bi(cyclohexane)

3.3. Mesomorphic Properties

DSC thermograms and some selected POM micrographs showing LC morphologies are presented in Figures 8 and 9, respectively. These were obtained during the first heating ramp. The only thermal phenomenon detected during DSC analysis between 20 °C and 100 °C was the nematic to isotropic transition. Data show that the nematic to isotropic transition temperatures (T_{NI}) increase according to the following order: A (67.5 °C), B (69 °C), D (72 °C) and C (80.8 °C). Distinct changes of T_{NI} were observed in Figure 8, related to a large number of LC molecules with various chemical structures, which are present in EOL-LCD screens. These LCD screens originated from different manufacturers, were fabricated at varied production dates and presented distinct dimensions. Each manufacturer uses a specific original LC mixture according to the type of screen and the technology to be developed. DSC results were found to be reasonably consistent with the findings from POM observations. Upon further heating above T_{NI}, the compounds reach the isotropic state, and the field of view turns into a transparent state. Thus, the extracted samples exhibited stable LC behavior with different mesomorphic temperature ranges.

Figure 9 shows micrographs obtained from three different temperatures for each sample: (1) at room temperature around 25 °C corresponding to the nematic state, (2) at a temperature at the beginning of the transition between nematic and isotropic states (or clearing point) and finally, (3) at a temperature in the isotropic phase. LC morphologies of these samples are different from each other. Thus, in addition to the changes in T_{NI}, the range of clearing points is different according to each sample (see Table 4). Among these four samples, C-LCs exhibited the widest mesomorphic temperature range and the highest clearing point, but its mesophase textures were poorly organized with less developed Schlieren texture. This seems to be consistent with the chemical structures observed for this sample. Indeed, in the C-LC mixture, "-HC=CH-" and "C=O" (or C≡N) bonds were not detected. The different chemical groups play an important role in influencing the mesomorphic behavior of LCs since they facilitate the formation of mesophases. Some authors indicate, for instance, that the alkoxy terminal groups can increase clearing points or promote the mesophase state of some organic molecules, whereas the absence of this group in the same alkyl chain decreases this mesomorphic state [30]. The chemical composition of these mixtures considerably influences their morphologies obtained by POM.

Table 4. Mesomorphic and thermal properties of LCs extracted from LC mixtures of EOL-LCDs A, B, C and D.

Sample	T_{NI} (°C)	Clearing Point Range T (°C) [ΔT]
A	67.5	65.8–72 [6.2 °C]
B	69	70.8–73.4 [2.6 °C]
C	80.8	79–88.2 [9.2 °C]
D	72	72.2–74 [1.8 °C]

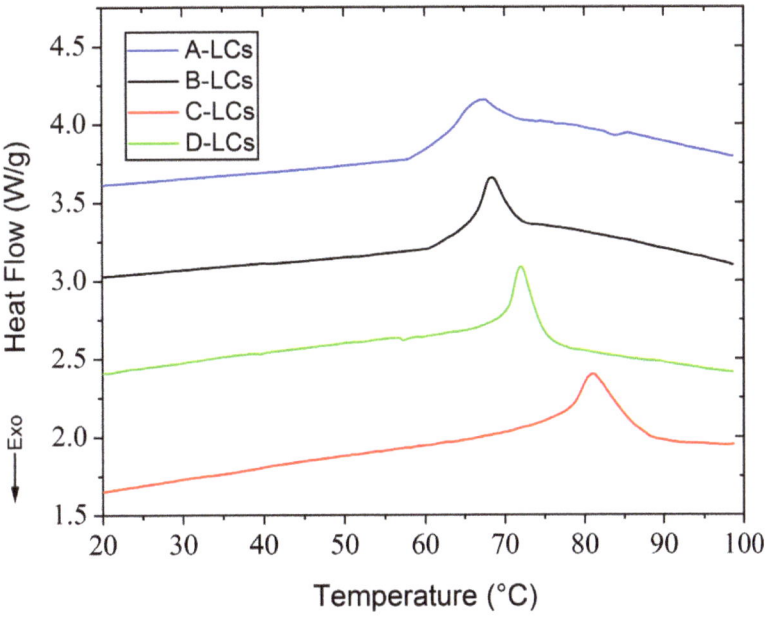

Figure 8. DSC results from LC mixtures of EOL-LCDs A, B, C and D.

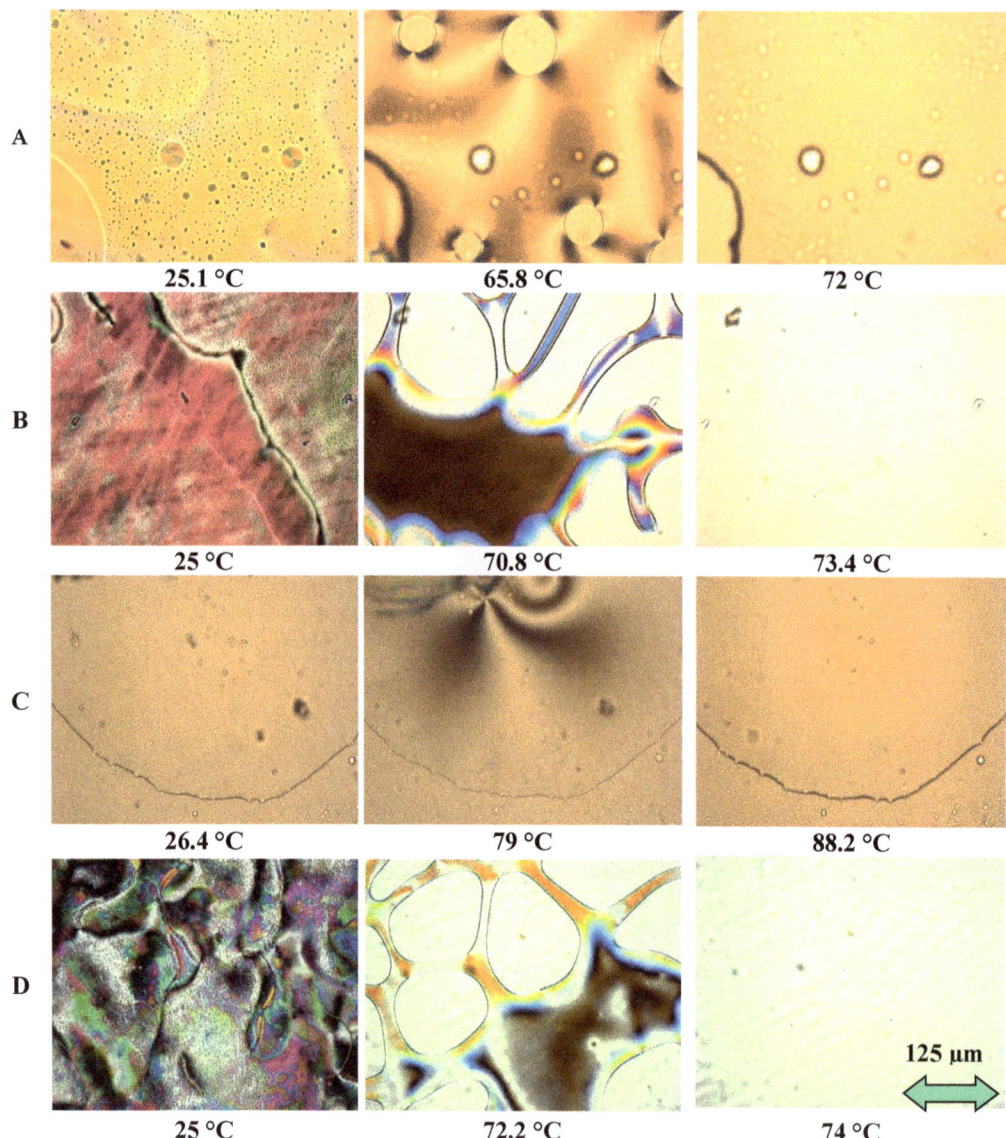

Figure 9. POM results from LC mixtures of EOL-LCDs A, B, C and D.

3.4. Thermogravimetric Analysis

Results from TGA analysis of the LC mixtures in an oxidative atmosphere (air) are shown in Figure 10, from which it can be concluded that all compounds have good thermal stability up to 150 °C in the presence of oxygen. Three of these samples, A, B and D, exhibit the same thermal behavior with only one degradation step around 150 °C. Sample C shows two degradation steps, the first at 150 °C identical to the previous three cases and the second at 250 °C characterized by a low slip.

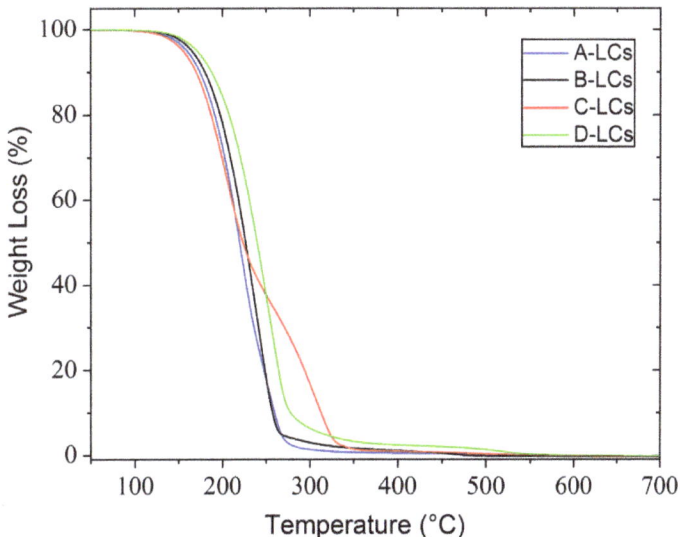

Figure 10. TGA results from LC mixtures of EOL-LCDs A, B, C and D.

3.5. Economic Interest of Extracted LCs

An economic balance was established to estimate the financial profitability of the recovery of LCs extracted from EOL-LCD panels. This calculation was based on 2500 tons of panels per year or 680 panels per day for 220 working days per year. This number corresponds to a rate of 97 panels per hour (one panel in 37 s) for a working week of five days and seven hours of daily activity. The calculation also considered the panel size, which is required for the estimation of LCs quantities to be extracted. The most representative size of LCD panels in the current EOF-LCD waste stream is 35 inches.

3.5.1. Evaluation of the Amount of Liquid Crystals

The weight of the LC residues depends on the space between the two plates of glass or the thickness (T) of LCs in the layers. This thickness can vary between 3 µm and 5 µm. LCs are distributed over the entire inner face of the glass plate (active area, S). Finally, the specific gravity (d) of the LCs varies from 0.97 to 1 g/cm^3. Hence, the following simple formula provides the theoretical mass of LCs:

$$\text{Mass of extracted LCs (g)} = T \times S \times d. \qquad (1)$$

The expected quantities of LCs are provided in Table 5. Applying a 60% recovery rate for a manual process, 678 g of LC material can be obtained considering 680 panels of 35-inch diameter per day. The recovery techniques must be implemented as an efficient process.

Table 5. Evaluation of mass of extracted LC from EOL-LCD (panels of 35-inch diameter).

LC Interlayer Thickness	3 µm	4 µm	5 µm
Specific gravity	1 g/cm^3	1 g/cm^3	1 g/cm^3
Theoretical mass (one panel)	1.01 g	1.35 g	1.69 g
« Real » mass (one panel, for 60% of recovery rate)	0.61 g	0.81 g	1.01 g
Total per day (680 panels)	415 g	551 g	678 g

3.5.2. Economic Benefits and Reuse Fields

The average price of LC mixtures sold for LCD applications varies between EUR 10 per gram and approximately EUR 100 per gram in the current market. As the use of LCDs continues to rise significantly, the economic stakes are huge, fully justifying the development of a LC recovery process. The LCs extracted from EOL-LCD panels have many potential uses:

(a) LCD flat panels

The extracted molecule LCs can be recycled in the production of LCD flat panels if their original electro-optical properties are maintained. It is conceivable in this case to separate and purify the LC molecules in order to increase the recovery profitability. For example, the Industrial Technology Research Institute (ITRI) has recovered and reused 100 tons of LC in order to assist Taiwanese LCD developers [18].

(b) Their use in PDLC systems

Purified LC molecules from EOL-LCD panels can be integrated into PDLC systems (Polymer Dispersed Liquid Crystals). By making a suitable choice of components of the mixture, it is possible to obtain a transparent film when the system is subjected to a sufficient electric voltage ("on" state) and opaque in the absence of voltage ("off" state). These materials are of interest for their many applications, especially in the field of electroactive glazing and display devices [31–34].

(c) LCs recycled as lubricants

Recent studies conducted at the Fraunhofer Institute for Mechanics of Materials (IMM) in Freiburg, Germany, have provided surprising results. While the friction force remains constant with oil, when using an LC layer, the friction forces start to decrease to almost zero after a certain time [35,36]. This application opens new perspectives for using LCs, whose use was mainly oriented in the electronics and telecommunications fields. Moreover, it would appear that the use of lower-quality LCs is as effective as using pure LCs in LCD panels.

(d) Recycled liquid crystals and thermal effects (thermochromic molecules)

The majority of LC applications utilize their sensitivity to temperature. Extracted LCs could also be used in temperature indicators (food packaging, medical sector), document security domains (passports, banknotes, designer clothes labels), battery testers and other voltage measuring devices, medical thermography, detection of radiation and thermal mapping.

4. Conclusions

EOL-LCD panels contain several valuable materials such as plastics, metals and electronics that account for more than 90% by weight of all components of LCD panels. Even if LC molecules represent less than 0.1% of all components by weight, their extraction and the possibility of introducing them in new processes have gained real technological and ecological interest. Indeed, this work presents the possibility of removing LC molecules from EOL-LCDs by a simple mechanical method without the use of organic solvents. The results show that all extracted LCs retain good optical, chemical and physical properties despite the aging effects of the panels. The residue of LCs obtained from mechanical extraction seems to be a good candidate material for further optical and electro-optical applications. A high degree of purity of the extracted LC molecules was observed, principally by NMR analysis, revealing the presence of chemical groups typical of nematic LC structures. Moreover, POM and DSC studies also corroborate the findings of NMR. In particular, only one nematic isotropic phase transition was found, indicating the existence of a homogeneous LC phase.

One of the aims of this work was to enhance the yield of collected LC molecules from used panels, purifying these products and looking forward to finding some interesting applications. These LC molecules might be found in new flat LCD panels after the purification of LC concentrate, in polymer/LC systems, in thermal applications, i.e., as temperature indicators, or they are used as lubricant to decrease friction forces. In order to bridge the gap between academic studies and industrial applications, continuous research and development work must be performed for all cases mentioned here. Studies will be undertaken in our laboratory to further investigate the challenging issue of reusing recycled LCs as mixtures.

Author Contributions: Conceptualization, U.M.; methodology, I.M., G.-J.F.T. and U.M.; validation, Z.B. and F.D.; investigation, I.M., G.-J.F.T., A.B. and Y.D.; data curation, I.M. and G.-J.F.T.; writing—original draft preparation, I.M., U.M., A.B., P.S. and C.F.; writing—review and editing. All authors have read and agreed to the published version of the manuscript.

Funding: This research was funded by the French Agency for Environment and Energy (ADEME), the University of Lille, the Région Hauts-de-France (FEDER) and the ENVIE^{2}E Company. The APC was funded by MDPI and by the University of Lille/France.

Institutional Review Board Statement: Not applicable.

Informed Consent Statement: Not applicable.

Data Availability Statement: The dataset presented in this study is available in this article.

Acknowledgments: The authors acknowledge financial support from the French Agency for Environment and Energy (ADEME), the University of Lille, the Région Hauts-de-France (FEDER), the ENVIE^{2}E Company, and MDPI.

Conflicts of Interest: All authors declare no conflict of interest.

Sample Availability: Samples of the compounds are currently under use for other applications.

References

1. Directive 2002/96/EC of the European Parliament and of the council of 27 January 2003 on waste electrical and electronic equipment (WEEE). *Off. J. Eur. Union* **2003**, *L37/24*, 12–25.
2. Directive 2002/95/EC of the European Parliament and of the council of 27 January 2003 on the restriction of the use of certain hazardous substances in electrical and electronic equipment. *Off. J. Eur. Union* **2003**, *L174/88*, 19–23.
3. Décret n 2005-829 du 20 juillet 2005 relatif à la composition des équipements électriques et électroniques et à l'élimination des déchets issus de ces équipements. *J. Off. République Française* **2005**, *169*, 8.
4. Décret n 2012-617 du 2 mai 2012 relatif à la gestion des déchets de piles et accumulateurs et d'équipements électriques et électroniques. *J. Off. République Française* **2012**, *105*, 19–72.
5. Décret n 2014-928 du 19 août 2014 relatif aux déchets d'équipements électriques et électroniques et aux équipements électriques et électroniques usagés. *J. Off. République Française* **2014**, *193*, 10–56.
6. European Commission. Circular Economy Action Plan. Available online: https://ec.europa.eu/environment/strategy/circular-economy-action-plan_en (accessed on 19 April 2021).
7. Conseil Supérieur de L'audiovisuel. L'équipement Audiovisuel des Foyers au 1er Semestre 2018. Available online: https://www.csa.fr/Informer/Collections-du-CSA/Panorama-Toutes-les-etudes-liees-a-l-ecosysteme-audiovisuel/Les-observatoires-de-l-equipement-audiovisuel/L-equipement-audiovisuel-des-foyers-au-1er-semestre-2018 (accessed on 19 April 2021).
8. Singh, S. *Liquid Crystals Fundamentals*, 1st ed.; World Scientific Publishing Co. Pte. Ltd.: London, UK, 2002; ISBN 9810242506.
9. Brown, C.V. Physical Properties of Nematic Liquid Crystals. In *Handbook of Visual Display Technology*; Chen, J., Cranton, W., Fihn, M., Eds.; Springer: Berlin/Heidelberg, Germany, 2012; Volume 1–4, pp. 1343–1361. ISBN 9783540795674.
10. Collings, P.J.; Hird, M. *Introduction to Liquid Crystals: Chemistry and Physics*, 1st ed.; Taylor & Francis Ltd.: London, UK, 2017; ISBN 9781351989244.
11. Yang, D.K.; Wu, S.T. *Fundamentals of Liquid Crystal Devices*, 1st ed.; John Wiley & Sons Ltd.: Chichester, UK, 2014; ISBN 9781118751992.
12. Goodship, V.; Stevels, A.; Huisman, J. *Waste Electrical and Electronic Equipment (WEEE) Handbook*, 2nd ed.; Woodhead Publishing (Elsevier): Cambridge, UK, 2019; ISBN 9780081021583.
13. Xenarc Tecnhologies. How The Technology of LCD Displays Works. Available online: https://www.xenarc.com/lcd-technology.html (accessed on 18 October 2021).
14. Ueberschaar, M.; Schlummer, M.; Jalalpoor, D.; Kaup, N.; Rotter, V.S. Potential and recycling strategies for LCD panels from WEEE. *Recycling* **2017**, *2*, 7. [CrossRef]
15. Kawamoto, H. The history of liquid-crystal display and its industry. In Proceedings of the 2012 Third IEEE HISTory of ELectrotechnology CONference (HISTELCON), Pavia, Italy, 5–7 September 2012; pp. 1–6.

16. Amato, A.; Beolchini, F. End of life liquid crystal displays recycling: A patent review. *J. Environ. Manage.* **2018**, *225*, 1–9. [CrossRef] [PubMed]
17. Hunt, A.J.; Clark, J.H.; Breeden, S.W.; Matharu, A.S.; Ellis, C.; Goodby, J.W.; Bottomley, J.A.; Cowling, S.J. Extraction of liquid crystals from flat panel display devices using both liquid and supercritical carbon dioxide. In Proceedings of the 11th International Supercritical Conference Proceedings, Barcelona, Spain, 4–7 May 2008; Volume 7, pp. 343–354.
18. ITRI. LCD Waste Recycling System-Circular Economy-Sustainable Environment-Innovations and Applications-Industrial Technology Research Institute. Available online: https://www.itri.org.tw/english/ListStyle.aspx?DisplayStyle=01_content&SiteID=1&MmmID=1037333532432522160&MGID=1037350654202216363 (accessed on 30 March 2021).
19. Izhar, S.; Yoshida, H.; Nishio, E.; Utsumi, Y.; Kakimori, N. Removal and recovery attempt of liquid crystal from waste LCD panels using subcritical water. *Waste Manag.* **2019**, *92*, 15–20. [CrossRef] [PubMed]
20. Fontana, D.; Forte, F.; De Carolis, R.; Grosso, M. Materials recovery from waste liquid crystal displays: A focus on indium. *Waste Manag.* **2015**, *45*, 325–333. [CrossRef] [PubMed]
21. Yoshida, H.; Izhar, S.; Nishio, E.; Utsumi, Y.; Kakimori, N.; Asghari, F.S. Recovery of indium from TFT and CF glasses of LCD wastes using NaOH-enhanced sub-critical water. *J. Supercrit. Fluids* **2015**, *104*, 40–48. [CrossRef]
22. Souada, M.; Louage, C.; Doisy, J.Y.; Meunier, L.; Benderrag, A.; Ouddane, B.; Bellayer, S.; Nuns, N.; Traisnel, M.; Maschke, U. Extraction of indium-tin oxide from end-of-life LCD panels using ultrasound assisted acid leaching. *Ultrason. Sonochem.* **2018**, *40*, 929–936. [CrossRef] [PubMed]
23. Amato, A.; Becci, A.; Mariani, P.; Carducci, F.; Ruello, M.L.; Monosi, S.; Giosuè, C.; Beolchini, F. End-of-life liquid crystal display recovery: Toward a zero-waste approach. *Appl. Sci.* **2019**, *9*, 2985. [CrossRef]
24. Zhang, K.; Li, B.; Wu, Y.; Wang, W.; Li, R.; Zhang, Y.N.; Zuo, T. Recycling of indium from waste LCD: A promising non-crushing leaching with the aid of ultrasonic wave. *Waste Manag.* **2017**, *64*, 236–243. [CrossRef] [PubMed]
25. Matharu, A.S.; Wu, Y. Liquid Crystal Displays: From Devices to Recycling. In *Issues in Environmental Science and Technology*; Hester, R.E., Harrison, R.M., Eds.; Royal Society of Chemistry: Cambridge, UK, 2009; pp. 180–211. ISBN 978-0-85404-112-1.
26. Barrera, A.; Binet, C.; Dubois, F.; Hébert, P.A.; Supiot, P.; Foissac, C.; Maschke, U. Dielectric spectroscopy analysis of liquid crystals recovered from end-of-life liquid crystal displays. *Molecules* **2021**, *26*, 2873. [CrossRef] [PubMed]
27. Maschke, U.; Moundoungou, I.; Fossi-Tabieguia, G.J. Method for Extracting the Liquid Crystals Contained in an Element that Comprises a First Support and a Second Support-Associated Device. Patent No. EP3111276 (A1), 1 April 2017.
28. Bezborodov, V.S.; Petrov, V.F.; Lapanik, V.I. Liquid crystalline oxygen containing heterocyclic derivatives. *Liq. Cryst.* **2006**, *20*, 785–796. [CrossRef]
29. Pakiari, A.H.; Aazami, S.M.; Ghanadzadeh, A. Electronic interactions of typical liquid crystal molecules with typical contacted species generated from the surface of different materials. *J. Mol. Liq.* **2008**, *139*, 8–13. [CrossRef]
30. Mandle, R.J.; Bevis, E.; Goodby, J.W. Phase Structures of Nematic Liquid Crystals. In *Handbook of Liquid Crystals: Physical Properties and Phase Behaviour of Liquid Crystals*; Wiley-VCH Verlag GmbH & Co.: Weinheim, Germany, 2014; pp. 1–27.
31. Bouchakour, M.; Derouiche, Y.; Bouberka, Z.; Beyens, C.; Supiot, P.; Dubois, F.; Riahi, F.; Maschke, U. Electron Beam Curing of Monomer/Liquid Crystal Blends. In *Polymer-Modified Liquid Crystals*; Dierking, I., Ed.; Royal Society of Chemistry: Cambridge, UK, 2019; pp. 45–60. ISBN 9781782629825.
32. Bouchakour, M.; Derouiche, Y.; Bouberka, Z.; Beyens, C.; Mechernène, L.; Riahi, F.; Maschke, U. Optical properties of electron beam- and UV-cured polypropyleneglycoldiacrylate/liquid crystal E7 systems. *Liq. Cryst.* **2015**, *42*, 1527–1536. [CrossRef]
33. Jain, A.K.; Deshmukh, R.R. An Overview of Polymer-Dispersed Liquid Crystals Composite Films and Their Applications. In *Liquid Crystals and Display Technology*; IntechOpen: London, UK, 2020; pp. 11–78. ISBN 978-1-78985-368-1.
34. Saeed, M.H.; Zhang, S.; Cao, Y.; Zhou, L.; Hu, J.; Muhammad, I.; Xiao, J.; Zhang, L.; Yang, H. Recent advances in the polymer dispersed liquid crystal composite and its applications. *Molecules* **2020**, *25*, 5510. [CrossRef] [PubMed]
35. Sengupta, A.; Schulz, B.; Ouskova, E.; Bahr, C. Functionalization of microfluidic devices for investigation of liquid crystal flows. *Microfluid Nanofluid* **2012**, *13*, 941–955. [CrossRef]
36. Yang, J.; Yuan, Y.; Li, K.; Amann, T.; Wang, C.; Yuan, C.; Neville, A. Ultralow friction of 5CB liquid crystal on steel surfaces using a 1,3-diketone additive. *Wear* **2021**, *480*, 203934. [CrossRef]

Article

Direct Ink Writing of Anisotropic Luminescent Materials

Mattia Sabadin [1,2], Jeroen A. H. P. Sol [2] and Michael G. Debije [2,*]

1. Materials and Process Engineering, Department of Engineering and Architecture, Università degli studi di Trieste (UniTS), 34127 Trieste, Italy
2. Laboratory of Stimuli-Responsive Functional Materials and Devices (SFD), Department of Chemical Engineering and Chemistry, Eindhoven University of Technology (TU/e), 5600 MB Eindhoven, The Netherlands
* Correspondence: m.g.debije@tue.nl

Abstract: Luminescent solar concentrators are relatively inexpensive devices proposed to collect, convert, and redirect incident (sun)light for a variety of potential applications. In this work, dichroic dyes are embedded in a liquid crystal elastomer matrix and used as feedstock for direct ink writing. Direct ink writing is a promising and versatile application technique for arbitrarily aligning the dichroic dyes over glass and poly(methyl methacrylate) lightguide surfaces. The resulting prints display anisotropic edge emissions, and suggest usage as striking visual objects, combining localized color and intensity variations when viewed through a polarizer.

Keywords: luminescent solar concentrator; liquid crystal elastomer; direct ink writing; fluorescent dye

1. Introduction

Luminescent solar concentrators (LSCs) are devices for collecting, converting, and redistributing sunlight [1–7], and by combining a straightforward architecture with the ability to control and modify light, have been proposed for use in a wide variety of applications, including as electricity generators [8], for catalyzing chemical reactions [9], enhancing horticultural production [10,11], as switchable "smart" windows [12], and for hydrogen production [13], among others [6]. The basic design consists of a transparent lightguide plate filled or topped with a fluorescent material. Incident (sun)light absorbed by the fluorophore is re-emitted at longer wavelengths, a fraction of this emitted light becoming trapped within the higher refractive index lightguide by total internal reflection, escaping primarily from the edges of the lightguide.

Many organic fluorescent molecules have extended structures with absorption and emission transition dipole moments more-or-less coincident with their molecular axis. This means that, if macroscopically aligned, the dyes absorb specific linear polarizations in preference to others. Likewise, emission of light is similarly expressed more strongly in directions favoring emission perpendicular to the transition dipole moment axes than parallel [14].

Considerable efforts have been directed towards improving the efficiency of LSC devices. One approach involves aligning the luminophore molecules of the LSC homeotropically to reduce photon losses through the surface(s) of the lightguide after the emission process and so increasing the efficiency [15–17]. When a suitably dichroic dye is used, it is also possible to direct the emission by aligning the dye planar to the lightguide surface [16,18–20]: by directing a significant fraction of the emitted light towards two rather than four edges, it will be possible to deploy two rather than four photovoltaics on the most radiant edges, reducing the overall cost of the device.

Nematic liquid crystals (LCs) have been used as a host for aligning dichroic organic luminophores due to their self-assembling ability [16,18,19]. Previous work has demonstrated deposition of oriented liquid crystalline elastomers (LCEs) via the additive manufacture

technique of direct ink writing (DIW), a microextrusion technique wherein it has been shown that printed LC inks have alignment following the print path [21–23]. These alignments have been further verified using the dichroic nature of embedded dyes [24,25]. In this work, we deposit an LCE material embedded with organic fluorescent dyes on two different substrates, glass and PMMA, to generate anisotropic LSC devices. If printing of aligned fluorescent dyes is possible, this would allow for complex optical structures with dye alignments varying over the surface of the device, which could find use in not only LSCs, but also security features or advanced light control elements.

2. Materials and Methods

The basic procedure for producing the LCE involves a thiol-Michael addition reaction described previously [23]. In summary, the LC diacrylate mesogens 2-methyl-1,4-phenylene bis(4-((6-(acryloyloxy)hexyl)oxy)benzoate) ("RM82") and 2-methyl-1,4-phenylene bis(4-(3-(acryloyloxy)propoxy)benzoate) ("RM257", both from Merck KGaA) are added in a 1:1:2 molar ratio with the chain extender 2,2′-(ethylenedioxy)diethanethiol (TCI Europe N.V.), and dissolved in dichloromethane (Biosolve) at 4 mL solvent per mg of reactants, in a 250 mL flask. The mix is stirred at room temperature for approximately 90 min before catalyst dimethylphenylphosphine (Sigma-Aldrich) is added. The reaction is rapid [26]; but we left it for about 1 h to ensure the reaction went to completion.

The thermal polymerization inhibitor 2,6-di-*tert*-butyl-4-methylphenol (0.05 wt%), 2 wt% photoinitiator phenylbis(2,4,6-trimethylbenzoyl)phosphine oxide and 0.05-0.1 wt% of the organic dye DFSB-K160 ("K160", Risk Reactor Inc.) or 0.05-0.2 wt% 4-(dicyanomethylene)-2-methyl-6-(4-dimethylaminostyryl)-4*H*-pyran ("DCM", Sigma-Aldrich) was then added to the solution and mixed 30 min at room temperature. The mix is transferred to a PTFE evaporation dish on a hot plate and left for 90 min at 50 °C, and then placed in a vacuum oven for 30 min at 80 °C to evaporate the remaining solvent. The final material is a sticky LC oligomer, suitable as DIW feedstock.

Ink deposition is accomplished using a Hyrel EHR 3D printer equipped with a TAM-15 high-operating temperature syringe extrusion head with a nozzle diameter of 0.335 mm (Fisnar QuantX Micron-S Red). Squares of 2×2 cm^2, and 2.5×2.5 cm^2 are printed on PMMA ($50 \times 50 \times 5$ mm^3) or glass ($30 \times 30 \times 1$ mm^3) substrates at printing speeds between 700 to 1000 mm min^{-1}. Bed temperature was RT for the glass substrates and fixed at 30 °C for the PMMA substrates, while the temperature of the syringe is set at 10 °C below the nematic-isotropic transition, T_{NI}, of the ink (usually around 70 °C, sample DSC data may be found in the Supplementary Materials (SM) as Figure S1). After printing, which takes about 1 min, the samples are immediately illuminated with a high-intensity UV lamp (Excelitas EXFO Omnicure S2000) for 15 min on the printed side and 15 min through the rear side to polymerize and form the crosslinked elastomer.

NMR spectra (example spectrum seen as Figure S2 in the Supplementary Materials) were recorded with a Bruker Avance III HD 400 MHz in chloroform-d (purchased from Sigma-Aldrich Inc., 99.8 atom % D, 0.03% v/v tetramethylsilane). Average chain length was determined by comparing the ratio of acrylate to mesogenic core signal [23]. Thermal behavior of the inks was investigated with a TA Instruments DSC Q2000. Polarized absorbances of the samples were recorded using a PerkinElmer Lambda 750 UV-vis-NIR spectrophotometer equipped with an integrating sphere detector and rotating linear polarizer. Surface profiles and thickness of the prints were evaluated using a Sensofar S neox 3D optical profiler. To measure edge emission, samples were illuminated by a 300 W solar simulator (Lot-Oriel) and output from sample edges was measured using a SLMS 1050 integrating sphere (Labsphere) equipped with a diode array detector (RPS900, International Light). A correction was made for the small polarization anisotropy (~10%) of the solar simulator. Internal efficiencies [27,28] were calculated as:

$$\eta_{int} = \text{\# photons emitted} \Big/ \text{\# photons absorbed} \tag{1}$$

3. Results and Discussion

Two commercial fluorescent dyes were used in this work, the orange laser dye DCM and lime-emitting K160. These two dyes have been used previously in spin-coated liquid crystal thin films [18]. The dye-doped oligomeric inks were DIW on both glass and PMMA substrates and photopolymerized to form the elastomers. Examples of K160 samples printed on the two different substrates are shown in Figure 1a,b: no difference in LCE adhesion to either substrate was observed.

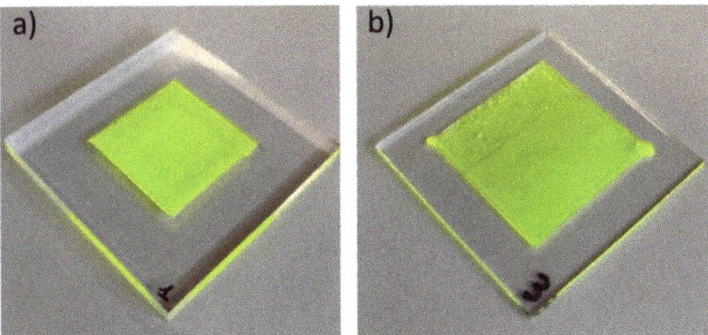

Figure 1. Photographs of printed squares of the LC oligomer containing 0.05 wt% K160 on (**a**) PMMA and (**b**) glass.

The dichroic order parameter, S, a common parameter to describe LC and dye alignments [3,12], of the fluorescent dye in the LC host is determined by measuring the absorption of light polarized along (A_{par}) and perpendicular (A_{per}) the direction of the liquid crystal director using:

$$S = \frac{A_{par} - A_{per}}{A_{par} + 2A_{per}} \quad (2)$$

The absorbance values are typically measured in a spectrophotometer equipped with a rotating polarizer (for a sample spectra and appearance of the film under polarized optical microscopy, see Figure S3 in the Supplementary Materials). From previous work using spin-coated LC films on pre-treated substrates, the S-values of the two dyes were roughly 0.25–0.45 for DCM and 0.4–0.57 for K160 [18]. For the extruded material, we generally find lower values for K160 of $S = 0.1$–0.3 on PMMA, depending on the resulting thickness of the applied layer, as seen in Figure 2, with thinner layers generally displaying greater order (all data may be found in the Supplementary Materials as Table S1). Two samples on glass showed significantly higher order, averaging 0.45. Three 0.2 wt% DCM samples between 75 and 108 μm thick averaged $S = 0.18$, so somewhat lower than K160, as previously noted [18]. Once printed and photopolymerized, further heating the samples reduced the order, but this loss was reversed upon cooling; see Figure S4 in the Supplementary Materials.

The reduced alignment in the DIW samples as compared to samples previously reported at least partly results from the increased thickness of the DIW written samples compared to spin coated samples; the DIW written samples rely entirely on the shear forces generated during printing for alignment, and typical printed LCEs themselves only have order parameters around $S = 0.3$ [21,29]. Additionally, the DIW samples demonstrate a more scattered appearance, suggesting reduced order within the deposited stripe. Regardless, the dyes do demonstrate a degree of anisotropic alignment. The advantage of using a DIW process is that aligned dyes may be deposited in almost any pattern at any location over the lightguide surface: this is considerably more complicated to achieve using spin coating, for example. It would also be possible to deposit different dyes over the same surface at discrete locations and deposition directions: while multicolor LSCs have been demonstrated [30–32], none have deployed aligned dye materials.

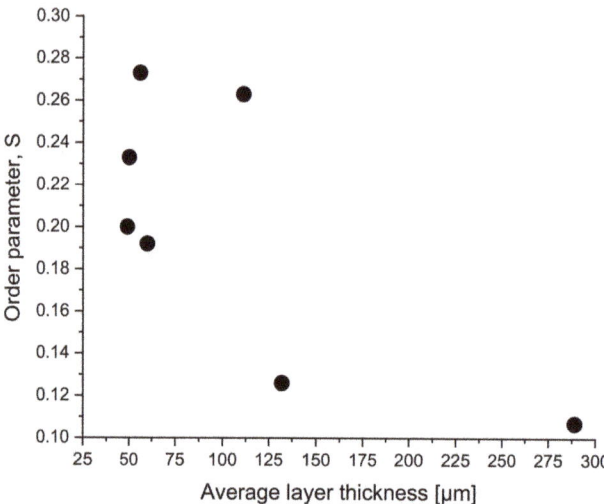

Figure 2. Derived order parameter, S, as a function of the thickness of the deposited layer for 0.05 wt% K160 deposited on PMMA.

The emission power output was recorded for each of the four edges of the samples. If the deposition direction of the DIW deposition is primarily parallel to the entry port of the integrating sphere it is called "parallel edge", while if it is perpendicular to the entry port it is called "perpendicular edge", similar to the earlier definition (see Figure 3a) [18]. Representative spectra obtained from the two edges for one DIW sample are shown in Figure 3b. The edge emission data and calculated internal efficiencies may be found in Table S1 in the Supplementary Materials.

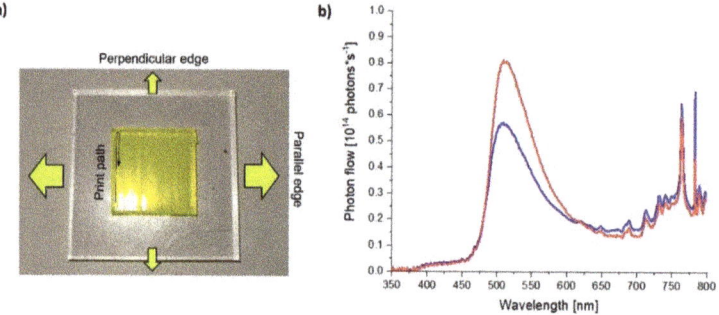

Figure 3. (**a**) Definition of perpendicular and parallel edge emissions. The direction of the original printing deposition is indicated, as well as the definitions of "perpendicular" and "parallel" edges. The green arrows represent the relative edge emissions from the two edges. (**b**) Representative comparison of edge emission spectra from the parallel (red line) and perpendicular (blue line) edges of a 0.05 wt% sample of K160 with $S = 0.26$. The spectral features >650 nm are measurement artifacts.

The resulting ratios between the parallel and perpendicular emissions for square samples on both PMMA and glass are plotted in Figure 4. Even though the order parameters obtained are only modest, for each sample the difference in the edge emission is quite discernable. In general, DIW of the oligomers produced considerably more scattering samples than spin-coated LCs. This manifests as additional scatter in the spectra, and readily visible in the somewhat "milky" appearance of the DIW samples. The printed squares also show surface texture from the rounded nature of the printed lines, also a source of scatter. As a result, the dichroism of emission for the DIW samples is considerably lower than that determined for the spin-coated samples [18].

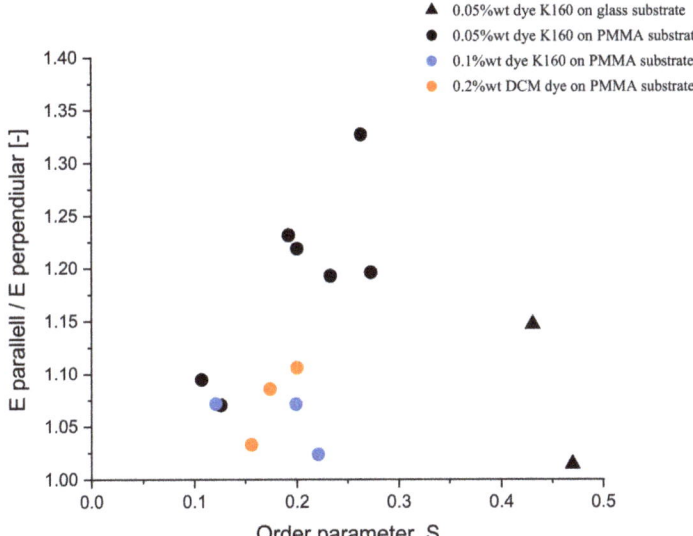

Figure 4. Emission ratio of parallel to perpendicular edges for K160 (0.05 wt% black, 0.1 wt% blue) and DCM (orange) samples on both glass (triangles) and PMMA (circles) substrates as a function of the order parameter, S.

Normally, one might expect better alignment should mean higher emission ratios but does not seem to be the case considering the glass samples (Figure 4 and Table S1 in the Supplementary Materials). We believe the diminished performance of the glass lightguides is a result of the increased frequency of dye layer encounters by light travelling through the lightguide by total internal reflection. Each encounter with the LCE/dye layers increases the probability of light scatter and surface losses, and potentially polarization randomization, and generation of more isotropic emission.

Despite the current modest order achieved in the printed dyes, there is interest in continuing to improve the quality of the alignment because of the potential application benefits. To demonstrate this potential, a sample was printed using DCM with different regions of the surface printed in different directions. Unfortunately, due to the surface structuring, the message integrated into the sample ("HI") was visible by eye. However, the difference between the text region and the background, printed with perpendicular orientations, was more dramatic when viewing the sample through a polarizer (Figure 5a depicts the sample viewed with transmission axis parallel to the printing direction of the text, and Figure 5b perpendicular, and Video S1 in the Supplementary Materials; a similar print using K160 may be seen in Figure S5). We are currently experimenting how to improve the surface finish, for instance with a solvent vapor exposure, the use of surfactant in the initial ink mixture, or through other means.

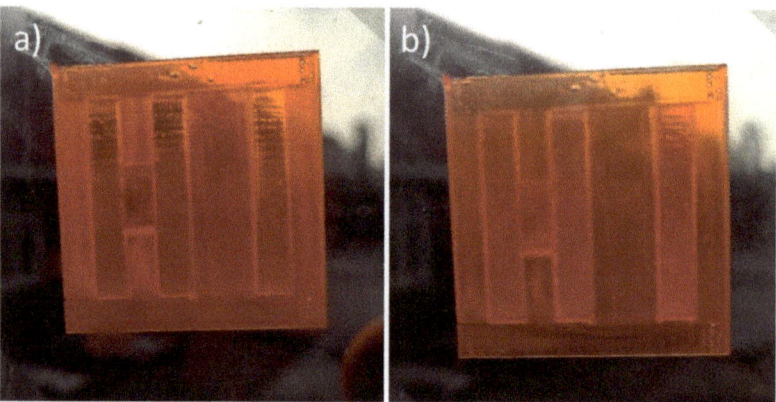

Figure 5. Direct ink-written LSC-type sample with LC oligomer containing DCM dye printed in two different directions viewed through a linear polarizer with absorption axis oriented (**a**) 0° and (**b**) 90° to the printing direction of the text.

By printing two differently colored inks on either side of a substrate, it is possible to depict images with different colors depending on the orientation of light polarization (Video S2 in the Supplementary Materials): this effect has been previously shown only using dyes that themselves aligned perpendicularly in a unpolymerized LC cell [33]. This printed sample also showed evidence of differently colored emission from perpendicular edges of the lightguide (see Figure S6 in the Supplementary Materials). Finally, circularly aligned samples were produced centered at different locations on the lightguide which could lead to high concentrations of light at the center or discrete edge regions of the plate (Figure S7) [20].

4. Conclusions

This work demonstrates DIW of two dichroic dyes embedded in a liquid crystal oligomer resulting in LSC-like devices with directional emission. Even though the order parameter, S, was modest compared to conventional samples deposited by spin coating, we still obtained positive edge emission ratios for each sample. Direct ink writing of transparent dye-doped LC elastomers presents a challenge, as the natural surface roughness of the prints and the thick layers result in reduced alignment and increased light scattering. Improving the printing parameters could allow application of the anisotropic emitting LSCs as security features, spectral converters and separators for use in agriculture, among others.

Supplementary Materials: The following supporting information can be downloaded at: https://www.mdpi.com/article/10.3390/cryst12111642/s1. Table S1: Material properties, ink compositions, deposition parameters, measured and derived properties of all the DIW LSC samples described in this work; Figure S1: Endothermic and exothermic DSC curves of a representative ink before inclusion of the dye; Figure S2: 1H NMR spectra of the synthesized ink; Figure S3: Polarized absorbance spectra of Coumarin yellow dye; Figure S4: Thermal cycling of K160 dye in LCE matrix; Figure S5: Direct ink written LSC-type sample; Figure S6: Photograph of PMMA lightguide and edge emission spectra from two perpendicular edges of the LSC device; Figure S7: DIW written K160 samples with the deposition done following an outwardly increasing spiral lightguide and edge centered; Video S1: Video depicting DIW sample using DCM dye viewed through a rotating polarizer; Video S2: Video depicting sample with DIW square containing K160 on top and DCM square deposited in opposite direction on bottom viewed through a rotating polarizer to show differences in apparent color.

Author Contributions: Conceptualization, M.G.D.; methodology, M.G.D. and J.A.H.P.S.; formal analysis, M.S.; investigation, M.S.; writing—original draft preparation, M.S.; writing—review and editing, M.G.D. and J.A.H.P.S.; All authors have read and agreed to the published version of the manuscript.

Funding: M.S. acknowledges the support of the European Union's Erasmus+ exchange program.

Data Availability Statement: Data available from the authors upon reasonable request.

Conflicts of Interest: The authors declare no conflict of interest.

References

1. Weber, W.H.; Lambe, J. Luminescent greenhouse collector for solar radiation. *Appl. Opt.* **1976**, *15*, 2299–2300. [CrossRef] [PubMed]
2. Goetzberger, A.; Greube, W. Solar energy conversion with fluorescent collectors. *Appl. Phys.* **1977**, *14*, 123–139. [CrossRef]
3. Debije, M.G.; Verbunt, P.P.C. Thirty years of luminescent solar concentrator research: Solar energy for the built environment. *Adv. Energy Mater.* **2012**, *2*, 12–35. [CrossRef]
4. McKenna, B.; Evans, R.C. Towards efficient spectral converters through materials design for luminescent solar devices. *Adv. Mater.* **2017**, *29*, 1606491. [CrossRef]
5. Roncali, J. Luminescent solar collectors: Quo vadis? *Adv. Energy Mater.* **2020**, *10*, 2001907. [CrossRef]
6. Papakonstantinou, I.; Portnoi, M.; Debije, M.G. The hidden potential of luminescent solar concentrators. *Adv. Energy Mater.* **2021**, *11*, 2002883. [CrossRef]
7. Ferreira, R.A.S.; Correia, S.F.H.; Monguzzi, A.; Liu, X.; Meinardi, F. Spectral converters for photovoltaics—What's ahead. *Mater. Today* **2020**, *33*, 105–121.
8. Van Sark, W.G.J.H.M.; Barnham, K.W.J.; Slooff, L.H.; Chatten, A.J.; Büchtemann, A.; Meyer, A.; McCormack, S.J.; Koole, R.; Farrell, D.J.; Bose, R.; et al. Luminescent Solar Concentrators—A review of recent results. *Opt. Express* **2008**, *16*, 21773. [CrossRef]
9. Zondag, S.D.A.; Masson, T.M.; Debije, M.G.; Noël, T. The development of luminescent solar concentrator-based photomicroreactors: A cheap reactor enabling efficient solar-powered photochemistry. *Photochem. Photobiol. Sci.* **2022**, *21*, 705–717. [CrossRef]
10. Loik, M.E.; Carter, S.A.; Alers, G.; Wade, C.E.; Shugar, D.; Corrado, C.; Jokerst, D.; Kitayama, C. Wavelength-selective solar photovoltaic systems: Powering greenhouses for plant growth at the food-energy-water nexus. *Earth Future* **2017**, *5*, 1044–1053. [CrossRef]
11. Timmermans, G.H.; Hemming, S.; Baeza, E.; van Thoor, E.A.J.; Schenning, A.P.H.J.; Debije, M.G. Advanced optical materials for sunlight control in greenhouses. *Adv. Opt. Mater.* **2020**, *8*, 2000738. [CrossRef]
12. Debije, M.G. Solar energy collectors with tunable transmission. *Adv. Funct. Mater.* **2010**, *20*, 1498–1502. [CrossRef]
13. Panzeri, G.; Tatsi, E.; Griffini, G.; Magagnin, L. Luminescent solar concentrators for photoelectrochemical water splitting. *ACS Appl. Energy Mater.* **2020**, *3*, 1665–1671. [CrossRef]
14. Van Gurp, M.; van Heijnsbergen, T.; van Ginkel, G.; Levine, Y.K. Determination of transition moment directions in molecules of low symmetry using polarized fluorescence. II. Applications to pyranine, perylene, and DPH. *J. Chem. Phys.* **1989**, *90*, 4103–4111. [CrossRef]
15. Debije, M.G.; Van, M.-P.; Verbunt, P.P.C.; Broer, D.J.; Bastiaansen, C.W.M. The effect of an organic selectively-reflecting mirror on the performance of a luminescent solar concentrator. In Proceedings of the 24th European Photovoltaic Solar Energy Conference and Exhibition, Hamburg, Germany, 21–25 September 2009; pp. 373–376.
16. Mulder, C.L.; Reusswig, P.D.; Velázquez, A.M.; Kim, H.; Rotschild, C.; Baldo, M.A. Dye alignment in luminescent solar concentrators: II. Horizontal alignment for energy harvesting in linear polarizers. *Opt. Express* **2010**, *18*, A79–A90. [PubMed]
17. MacQueen, R.W.; Cheng, Y.Y.; Clady, R.G.C.R.; Schmidt, T.W. Towards an aligned luminophore solar concentrator. *Opt. Express* **2010**, *18*, A161–A166. [CrossRef]
18. Verbunt, P.P.C.; Kaiser, A.; Hermans, K.; Bastiaansen, C.W.M.; Broer, D.J.; Debije, M.G. Controlling light emission in luminescent solar concentrators through use of dye molecules aligned in a planar manner by liquid crystals. *Adv. Funct. Mater.* **2009**, *19*, 2714–2719. [CrossRef]
19. Verbunt, P.P.C.; de Jong, T.M.; de Boer, D.K.G.; Broer, D.J.; Debije, M.G. Anisotropic light emission from aligned luminophores. *Eur. Phys. J. Appl. Phys.* **2014**, *67*, 10201.
20. Bruijnaers, B.J.; Schenning, A.P.H.J.; Debije, M.G. Capture and concentration of light to a spot in plastic lightguides by circular luminophore arrangements. *Adv. Opt. Mater.* **2015**, *3*, 257–262. [CrossRef]
21. Kotikian, A.; Truby, R.L.; Boley, J.W.; White, T.J.; Lewis, J.A. 3D printing of liquid crystal elastomeric actuators with spatially programed nematic order. *Adv. Mater.* **2018**, *30*, 1706164.
22. Del Pozo, M.; Sol, J.A.H.P.; van Uden, S.H.P.; Peeketi, A.R.; Lugger, S.J.D.; Annabattula, R.K.; Schenning, A.P.H.J.; Debije, M.G. Patterned actuators via direct ink writing of liquid crystals. *ACS Appl. Mater. Interfaces* **2021**, *13*, 59381–59391. [CrossRef]
23. Del Pozo, M.; Sol, J.A.H.P.; Schenning, A.P.H.J.; Debije, M.G. 4D printing of liquid crystals: What's right for me? *Adv. Mater.* **2022**, *34*, 2104390. [CrossRef] [PubMed]
24. Gelebart, H.; Bride, M.M.; Schenning, A.P.H.J.; Bowman, C.N.; Broer, D.J. Photoresponsive fiber array: Toward mimicking the collective motion of cilia for transport applications. *Adv. Funct. Mater.* **2016**, *26*, 5322–5327. [CrossRef]
25. López-Valdeolivas, M.; Liu, D.; Broer, D.J.; Sánchez-Somolinos, C. Photoresponsive fiber array: Toward mimicking the collective motion of cilia for transport applications. *Macromol. Rapid Commun.* **2018**, *39*, 3–9.
26. Chan, J.W.; Hoyle, C.E.; Lowe, A.B.; Bowman, M. Nucleophile-initiated thiol-michael reactions: Effect of organocatalyst, thiol, and ene. *Macromolecules* **2010**, *43*, 6381–6388. [CrossRef]

27. Debije, M.G.; Evans, R.C.; Griffini, G. Laboratory protocols for measuring and reporting the performance of luminescent solar concentrators. *Energy Environ. Sci.* **2021**, *14*, 293–301.
28. Yang, C.; Atwater, H.A.; Baldo, M.A.; Baran, D.; Barile, C.J.; Barr, M.C.; Bates, M.; Bawendi, M.G.; Bergren, M.R.; Borhan, B.; et al. Consensus statement: Standardized reporting of power-producing luminescent solar concentrator performance. *Joule* **2022**, *6*, 8–15. [CrossRef]
29. Sol, J.A.H.P.; Douma, R.F.; Schenning, A.P.H.J.; Debije, M.G. 4D printed light-responsive patterned liquid crystal elastomer actuators using a single structural color ink. *Adv. Mater. Technol.* **2022**, 2200970. [CrossRef]
30. Albers, P.T.M.; Bastiaansen, C.W.M.; Debije, M.G. Dual waveguide patterned luminescent solar concentrators. *Sol. Energy* **2013**, *95*, 216–223.
31. Ter Schiphorst, J.; Cheng, M.L.M.K.H.Y.K.; van der Heijden, M.; Hageman, R.L.; Bugg, E.L.; Wagenaar, T.J.L.; Debije, M.G. Printed luminescent solar concentrators: Artistic renewable energy. *Energy Build.* **2019**, *207*, 109625. [CrossRef]
32. Renny, A.; Yang, C.; Anthony, R.; Lunt, R.R. Luminescent solar concentrator paintings: Connecting art and energy. *J. Chem. Educ.* **2018**, *95*, 1161–1166. [CrossRef]
33. Debije, M.G.; Menelaou, C.; Herz, L.M.; Schenning, A.P.H.J. Combining positive and negative dichroic fluorophores for advanced light management in luminescent solar concentrators. *Adv. Opt. Mater.* **2014**, *2*, 687–693. [CrossRef]

Article

Anisotropic Surface Formation Based on Brush-Coated Nickel-Doped Yttrium Oxide Film for Enhanced Electro-Optical Characteristics in Liquid Crystal Systems

Dong-Wook Lee [1], Da-Bin Yang [1], Dong-Hyun Kim [1], Jin-Young Oh [1], Yang Liu [2,*] and Dae-Shik Seo [1,*]

[1] IT Nano Electronic Device Laboratory, Department of Electrical and Electronic Engineering, Yonsei University, 50 Yonsei-ro, Seodaemun-gu, Seoul 03722, Korea
[2] College of Information Science and Technology, Donghua University, 2999 North Renmin Road, Songjiang District, Shanghai 201620, China
* Correspondence: liuyang@dhu.edu.cn (Y.L.); dsseo@yonsei.ac.kr (D.-S.S.)

Abstract: This paper introduces anisotropic nickel yttrium oxide (NYO) film formed by the brush coating technique. X-ray photoelectron spectroscopy confirmed well-formed NYO film after the curing process, and the morphology of the surface was investigated using atomic force microscopy. The shear stress driven from brush hair movements caused the nano/micro-grooved anisotropic surface structure of NYO. This anisotropic surface induced uniform liquid crystal (LC) alignment on the surface, which was confirmed by pre-tilt angle analysis and polarized optical microscopy. The contact angle measurements revealed an increase in hydrophilicity at higher temperature curing, which contributed to homogenous LC alignment. The NYO film achieved good optical transmittance and thermal stability as an LC alignment layer. In addition, the film demonstrated good electro-optical properties, stable switching, and significantly enhanced operating voltage performance in a twisted-nematic LC system. Therefore, we expect that this brush coating method can be applied to various inorganic materials to achieve an advanced LC alignment layer.

Keywords: brush coating; nickel yttrium oxide; surface morphology; liquid crystal; low voltage operation

Citation: Lee, D.-W.; Yang, D.-B.; Kim, D.-H.; Oh, J.-Y.; Liu, Y.; Seo, D.-S. Anisotropic Surface Formation Based on Brush-Coated Nickel-Doped Yttrium Oxide Film for Enhanced Electro-Optical Characteristics in Liquid Crystal Systems. *Crystals* **2022**, *12*, 1554. https://doi.org/10.3390/cryst12111554

Academic Editors: Zhenghong He and Yuriy Garbovskiy

Received: 13 October 2022
Accepted: 29 October 2022
Published: 31 October 2022

Publisher's Note: MDPI stays neutral with regard to jurisdictional claims in published maps and institutional affiliations.

Copyright: © 2022 by the authors. Licensee MDPI, Basel, Switzerland. This article is an open access article distributed under the terms and conditions of the Creative Commons Attribution (CC BY) license (https://creativecommons.org/licenses/by/4.0/).

1. Introduction

With the rapid development of surface engineering, interest in highly functional and economical devices is increasing in the optical, electronic, and display industries. Liquid crystals (LCs) are versatile materials that can be applied to many industries due to their unique properties, such as the intermediate state of liquid and solid, refractive anisotropy, and dielectric anisotropy [1–3]. In particular, LC displays (as a representative application of using LCs) have received significant attention due to their durability, excellent electro-optical (EO) properties, and high resolution [4–7]. For high-performance LC-based devices, uniform LC alignment should be achieved. This is because the LCs that uniformly aligned can control the light with high reliability, whereas irregularly distributed LCs cause the light leakage effect and unstable EO properties, which can deteriorate device performance [8,9]. Accordingly, LCs are located on the alignment layer, where the alignment state is affected by interactions between the LCs and the alignment layer.

Various treatments have been researched for the alignment layer to induce uniform LC alignment, including the rubbing method [10,11], ultraviolet photoalignment [12], oblique deposition [13], and plasma treatment [14]. Especially the method called rubbing has been extensively adopted in industry because of its simple and cost-effective properties. In this method, anisotropic microgrooves are produced on the alignment layer surface through contact with a high-speed rotating fabric roller to induce uniform LC alignment. However, this intense mechanical contact also causes cracks, local defects, and electrostatic problems, resulting in a breakdown of device performance [15].

Herein, we introduce the brush coating technique to induce uniform LC orientation on the alignment layer. This method is very convenient and can integrate the film formation process and treatment process for alignment layer, resulting in high throughput and cost-effectiveness. Moreover, this solution-based brush coating method can induce shear stress to the solution (by brush hair sweeping) and a retracing force on the deposited solution. We assumed that this shear stress could generate a directional distribution of the solution and the subsequent rapid heat application could form the anisotropic microgroove film surface with the sol-gel method. Nickel yttrium oxide (NYO) was used for the alignment layer due to its good dielectric characteristics, which means the potential of EO performance as an alignment layer [16,17]. The sol-gel-based brush coating method was adopted with a glass substrate, and the film curing temperature was varied. The film surface morphology was verified by atomic force microscopy (AFM). X-ray photoelectron spectroscopy (XPS) verified the well-formed NYO film state. Optical transparency of the layer was confirmed by ultraviolet-visible-near infrared (UV-vis-NIR) spectroscopy, and the atomic structural properties were verified by X-ray diffraction (XRD). In addition, contact angle investigation was conducted to verify the chemical affinity of the film surface. The LC orientation state was confirmed by polarized optical microscopy (POM) and pre-tilt angle analysis, and the EO characteristics of the film were examined in a twisted-nematic (TN) LC system.

2. Materials and Methods

To form the NYO film, 2 cm × 3 cm glass substrates were prepared. They were cleaned by the sonification method using isopropyl alcohol and acetone. The substrates were treated in each solvent for 10 min with subsequent drying with N_2 gas. The 0.1 M NYO solution was then produced by mixing nickel(II) chloride hydrate and yttrium(III) nitrate hexahydrate at a ratio of 1:9 in a 2-methoxyethanol solvent. The mixed solution was then stirred at 430 rpm for 2 h at 75 °C and then aged for at least 24 h. The used metal mateirals are for the sol-gel process, and thus the sol state of NYO is uniformly distributed in the solution. The prepared brush hair was saturated in the solution and the combing of the hair on the prepared glass substrate produced the film deposition process. Subsequently, the curing process was enacted under 70, 150, and 230 °C conditions. The curing made sol-gel transition, with decomposing and hardening the oxide material, and formed the NYO film.

Stoichiometric differences of the brush-coated NYO films (depending on curing temperature) were examined by XPS (K-alpha, Thermo Scientific, Waltham, MA, USA). A monochromatic Al X-ray source (Al Kα line:1486.6 eV) was used with a 12 kV/3 mA power source. The surface morphology information of the film surface was investigated using AFM (NX-10, Park Systems, Seoul, Korea) and corresponding line profile data. The dektakXT stylus profiler (Bruker, Billerica, MA, USA) was used for measuring the NYO film thickness with 2 μm radius tip. The thickness was measured to 243.32 nm. The optical transmittance of the film was then evaluated with UV-Vis-NIR spectroscopy (JASCO Corporation, V-650, Tokyo, Japan) using a wavelength range of 250–850 nm. Using the measurement result in air as a baseline, the transmittance of the glass substrate, the indium-tin-oxide (ITO)-coated glass substrate, and the NYO film coated glass substrate were measured, respectively. The surface chemical affinity of the film was examined through contact angle measurements using the sessile drop technique with deionized water and diiodomethane. A phoenix 300 surface angle analyzer and the IMAGE PRO 300 software were used for the analyses. The atomic structural properties of the film were examined by XRD (DMAX-IIIA, Rigaku, Tokyo, Japan) with a 2-theta range of 20–80°.

To verify the LC alignment state on the brush-coated NYO film, anti-parallel (AP) cells were assembled by the brush-coated NYO films, which were cured at various temperatures. The cell gap was uniformly assembled to 60 μm, and then positive nematic LCs (IAN-5000XX T14, $T_{N \rightarrow I}$ = 81.8 °C, Δn = 0.111, ne = 1.595, no = 1.484; JNC) were injected into the cells by syringes via capillary force. The assembled LC cells were observed by POM (BXP 51, Olympus, Tokyo, Japan) to confirm the LC alignment state, and the pre-tilt of the LCs in the cell was investigated using the crystal rotation method (Autronic TBA 107).

Thermal stability to LC alignment was inspected by the annealing process and subsequent POM measurements. To examine the EO properties, a TN-LC cell was assembled with a uniform cell gap of 5 μm. The response time-transmittance (R-T) and voltage-transmittance (V-T) graphs were measured using the LCD evaluation system (LCMS-200) to confirm the switching and operating voltage characteristics.

3. Results and Discussion

As illustrated in Figure 1, brush hair movement during the brush coating process was expected to form directional NYO precursor distribution on the substrate via shear stress, which originated from the retracing force on the deposited bulk solution [18,19]. This sol state of the NYO could be transformed into a gel state during the curing process, forming a directional NYO film structure on the surface.

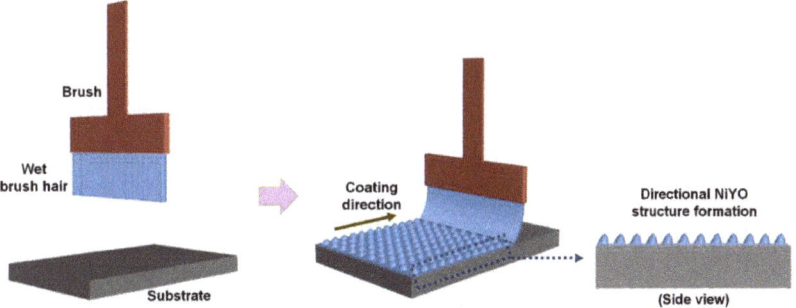

Figure 1. Brush coating process illustration and the expected structure of nickel yttrium oxide film after the curing process.

To verify the morphology of the NYO film, AFM measurements, and corresponding line profile analyses were conducted, as depicted in Figure 2. When the curing process progressed at 70 and 150 °C, some large lumps were measured on the surface that were attributed to large NYO particles between the brush hairs, although neither sample exhibited any distinct structural property with irregular morphology. The corresponding line profiles also denoted an irregular structure, which could not be perceived as anisotropic morphology. In contrast, the 230 °C cured sample exhibited an anisotropic structure that was aligned in a single direction, similar to the brush coating direction in the AFM result. In terms of line profile, the surface exhibited an anisotropic micro/nano-groove structure in which the height increased and decreased repeatedly according to the direction of brush coating. This anisotropic structure originated from the NYO precursors that were distributed during the brush coating process [20]. From these results, it was demonstrated that curing temperature is the important factor when forming an anisotropic NYO film structure while maintaining the directional property of the NYO precursors. In addition, the application of sufficiently high temperature should be guaranteed to ensure rapid transformation of the NYO sol state to a gel state. When the curing temperature was too low, sufficient transformation did not occur and unstable films were formed under the influence of residual solvent.

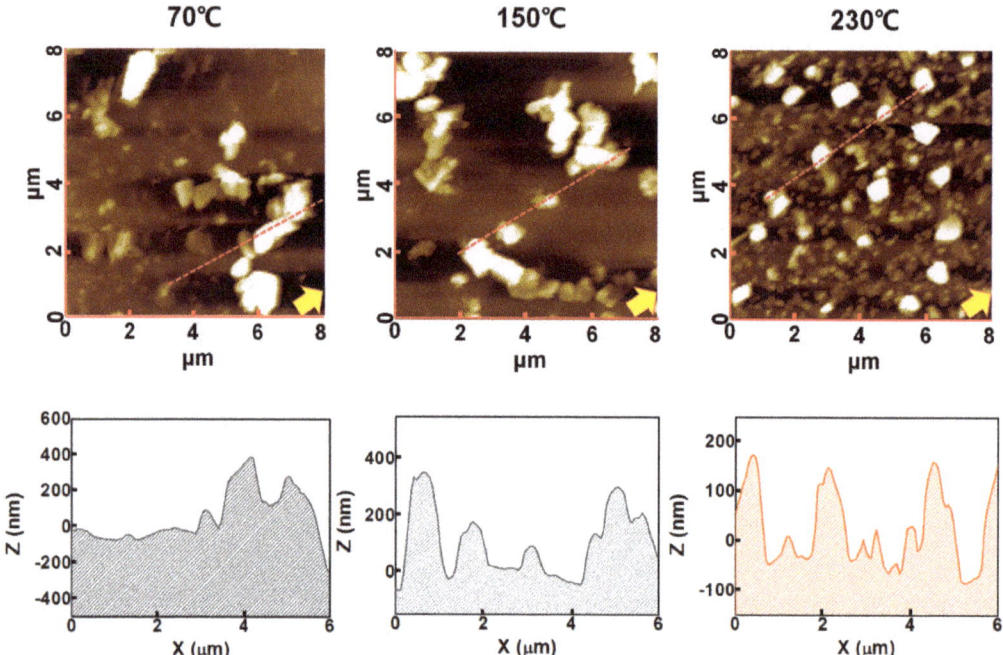

Figure 2. Atomic force microscopy images and corresponding line profiles of the brush−coated nickel yttrium oxide films cured at 70, 150, and 230 °C (The red dotted line corresponds to the line profile data). Direction of brush coating is denoted by a yellow arrow.

The stoichiometric difference between the 70 and 230 °C cured NYO films was investigated by XPS analysis, from which the Ni 2p, Y 3d, and O 1s core-level spectra were obtained (Figure 3). In the Ni 2p spectra, both samples revealed four sub-peaks, representing Ni 2p3/2, Ni 2p1/2, and two satellites. The 3/2 and 1/2 peaks were respectively centered at 854.46 and 872.13 eV for the 70 °C sample, compared to 854.38 and 871.85 eV for the 230 °C sample. The Y 3d spectra comprised two sub-peaks, which represented Y 3d5/2 and Y 3d3/2 in each for both samples. Each peak was centered at 157.89 and 159.84 eV for the 70 °C sample compared to 157.23 and 158.88 eV for the 230 °C sample. The O 1s spectra comprised two sub-peaks, each indicating metal-oxide bonding and oxygen vacancy. These peaks were centered in the ranges of 528.50–530.00 eV and 531.00–531.50 eV, respectively. The peak intensity increases of Ni 2p, Y 3d, and metal-oxide bonding as the curing temperature is increased from 70 to 230 °C indicated that thermal oxidation was well-progressed at the higher curing temperature [21]. This contributed to the formation of a stable NYO film during the curing process, although 70 °C curing was relatively insufficient for producing a stable oxidized NYO film structure, as confirmed by the AFM analysis.

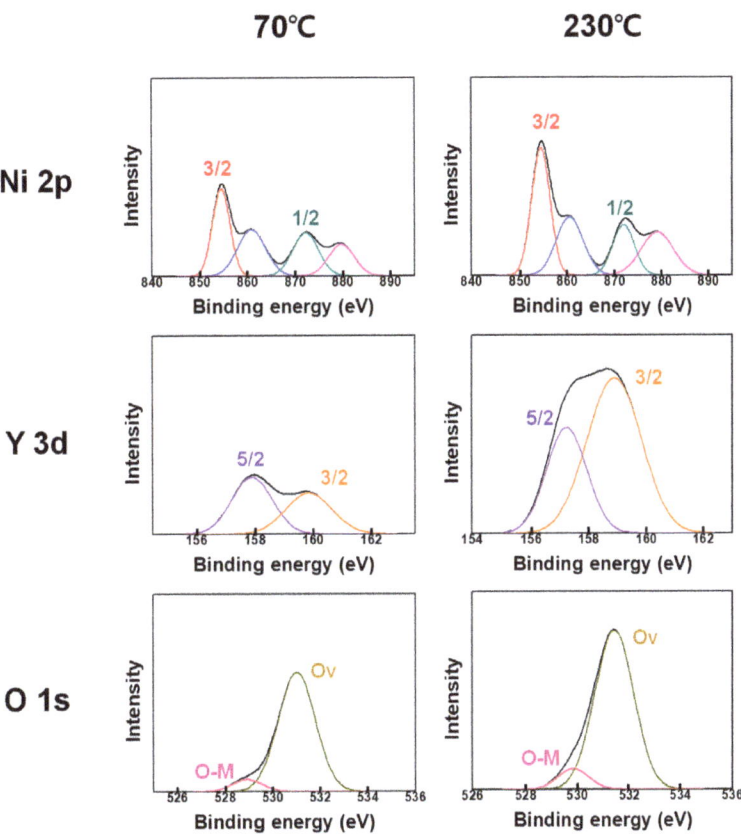

Figure 3. X-ray photoelectron spectroscopy results of Ni 2p, Y 3d, and O1s core levels obtained from brush-coated nickel yttrium oxide films which were cured at 70 and 230 °C.

To verify the adoptability of the brush-coated NYO film to the LC alignment layer, the film's optical transparency was measured (Figure 4). In addition, the transparency of plain and ITO coated glasses were obtained for comparison. The 70 and 150 °C cured films exhibited significantly lower transmittance curves compared to the 230 °C cured film. This confirmed that the optical properties of the 70 and 150 °C cured samples were deteriorated by the residual-solvent effect, whereas the 230 °C cured sample achieved a stable film state in terms of optical properties. The average transmittance of the 230 °C cured sample was 85.9% in the visible region (380–740 nm wavelength in this study). Considering that the corresponding values for the plain and ITO glasses were 86.5% and 82.3%, respectively, it was verified that the 230 °C cured brush-coated NYO film has the potential to be adopted as an LC alignment layer.

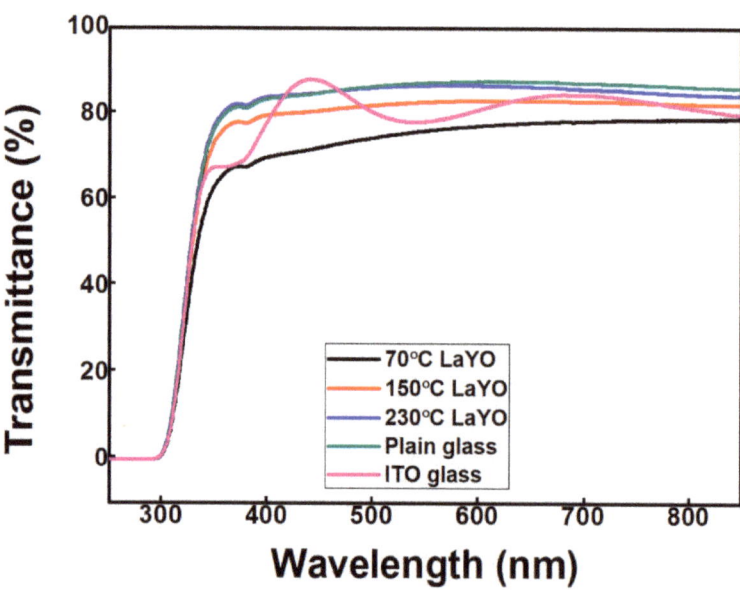

Figure 4. Optical transparency graphs acquired from the brush-coated nickel yttrium oxide films, plain glass, and indium-tin-oxide-coated glass.

The LC alignment state on the brush-coated NYO film was evaluated by POM measurements of AP LC cells assembled by the NYO films cured at 70, 150, and 230 °C. The corresponding POM results are represented in Figure 5a. The 70 and 150 °C cured sample-based LC cells revealed some defects and presented entirely yellowish images in POM when the polarizer and analyzer were vertically crossed. This signified instability of LC alignment and a light leakage effect of the LC cell, representing randomly distributed LCs on the NYO film. In contrast, the 230 °C cured sample-based LC cell exhibited a distinctly dark POM image without defects, indicating a uniform LC alignment state in the cell. The uniformly aligned LCs can guide the light unidirectionally, meaning that the polarized light from the polarizer could pass through the LC cell without any distortion. This light was blocked by the analyzer (placed after the LC cell), resulting in no light being observed in the POM measurement, as illustrated in Figure 5b.

The pre-tilt angle of LCs on the NYO alignment layer was investigated using the crystal rotation method via the oscillated transmittance curves of the LC cells, as depicted in Figure 6a [22,23]. The red line in the graph was obtained from the experimentally measured curve, whereas the blue line represented simulation data acquired from information on the LC cell gap and injected LCs in the cell. The 70 and 150 °C cured sample-based LC cells exhibited irregular experimental curves, which demonstrated a significant mismatch rate with the simulation data. This indicates an unstable LC distribution on the alignment layer; hence, the pre-tilt angle could not be obtained. In contrast, the 230 °C cured sample-based LC cell achieved a high match rate between the simulation and experimental data and the LC pre-tilt angle could be obtained (0.19°) with high reliability. This indicates a homogeneous LC alignment state on the NYO alignment layer. From the AFM, POM, and LC pre-tilt angle analysis, it was revealed that the directional anisotropic surface of the 230 °C cured NYO film could induce uniform and homogeneous LC alignment. With the nano/micro-grooved boundary of the surface, the LCs on the surface were constrained geometrically to the corresponding surface directionality, achieving uniform orientation [24–27]. The LCs have collective behavior characteristics and also fluidity, which are originated from the van der Waals forces between the LC molecules with accompanying elastic distortion. Hence, the surface LC molecules' orientation information is propagated to the LCs in the cell, achieving uniform LC alignment in the cell. The LC alignment on the brush-coated NYO film is illustrated in Figure 6b.

The chemical affinity of the brush-coated NYO film surface was analyzed using contact angle measurements, as shown in Figure 7a. Here, deionized water and diiodomethane were used and their contact angles were reduced from 57.0° to 38.8° and from 55.3° to 44.3°, respectively, as the film curing temperature was increased from 70 to 230 °C. The surface energies of the samples were calculated using the Owen–Wendt method, as depicted in Table 1 [28]. The increase in surface energy indicated that the hydrophilicity of the NYO surface increased as the curing temperature increased. This can contribute to the homogeneous alignment of LC molecules on the surface, and corresponds to the LC pre-tilt angle analysis. The crystallinity of the 230 °C cured brush-coated NYO film was also analyzed using XRD, as shown in Figure 7b. In the range of 20–80 2-theta degrees, no distinct peak was observed, indicating that the NYO film has an amorphous structure. Although the solution-processed oxide film generally exhibited an amorphous structure below 500 °C curing [29], the 230 °C cured brush-coated NYO film achieved uniform LC alignment on the surface. Therefore, this demonstrated that the amorphous structure of the alignment layer did not affect the uniform alignment of the LC molecules.

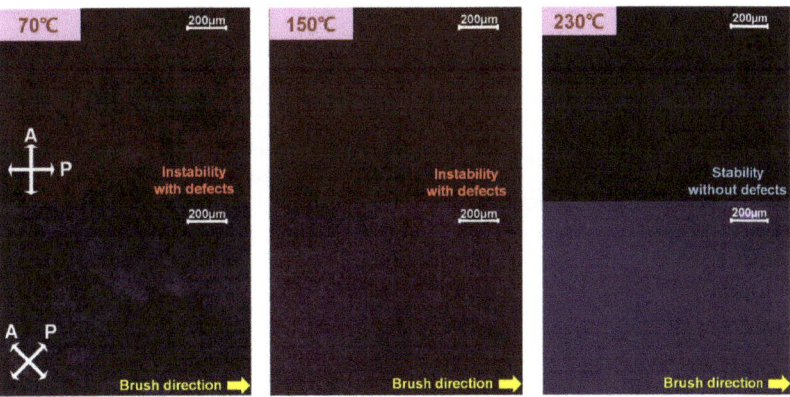

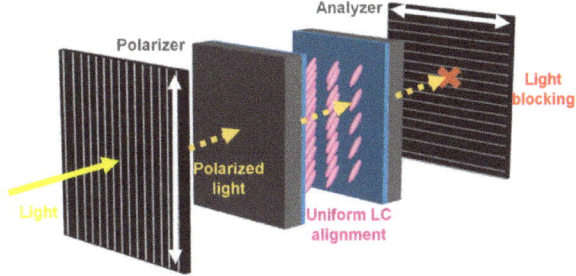

Figure 5. (**a**) Polarized optical microscopy results of anti-parallel liquid crystal (LC) cells made from brush-coated nickel yttrium oxide films cured at 70, 150, and 230 °C. The analyzer ("A") and polarizer ("P") directions are denoted by white arrows. (**b**) Schematic of light blocking process with uniformly aligned LCs in the cell.

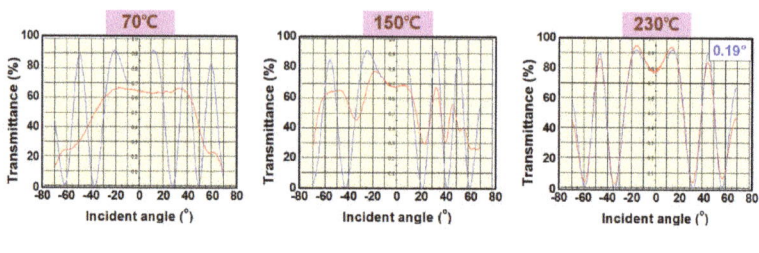

Figure 6. (**a**) Oscillated transmittance graphs measured from the anti-parallel LC cells based on the variously cured brush—coated nickel yttrium oxide films. The blue line denotes the simulation data, and the red line indicates the experimental data. (**b**) Schematic of the LC alignment on the anisotropic nickel yttrium oxide alignment layer.

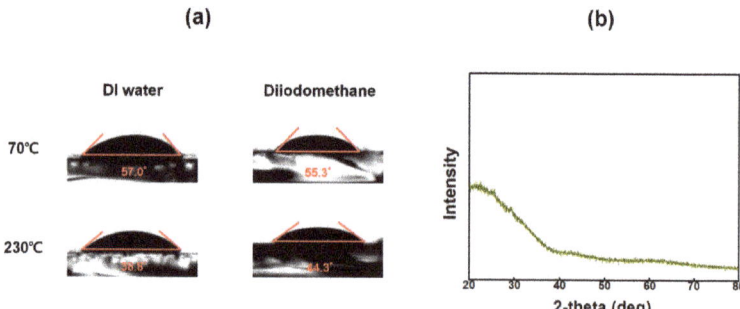

Figure 7. (**a**) Contact angle measurement results of the brush-coated nickel yttrium oxide films cured at 70 and 230 °C using deionized (DI) water and diiodomethane. (**b**) X-ray diffraction graph of the brush-coated nickel yttrium oxide film cured at 230 °C.

Table 1. Contact angles and surface energies of the brush-coated nickel yttrium oxide films cured at 70 and 230 °C.

Curing Temperature (°C)	Contact Angle (°)		Surface Energy (mJ/m^2)
	Deionized Water	Diiodomethane	
70	57.0	55.3	49.1
230	38.8	44.3	63.1

It is known that LC devices containing numerous switching components are subject to increasing temperature. Therefore, thermal stability of the alignment layer to uniform orientation of LCs is an significant factor in applications. Accordingly, the thermal endurance of the brush-coated NYO film to LC alignment was analyzed by the annealing process and POM measurements, as shown in Figure 8. Increasing temperature was applied from 90 to 180 °C at intervals of 30 °C. Uniform LC orientation state was verified by the POM results up to 150 °C. However, when the heat was increased to 180 °C, the POM result revealed defects, which indicated broken LC alignment. This result demonstrates the suitable thermal stability and potential of the NYO layer as an application to alignment layer of LCs compared to conventional PI layers [30].

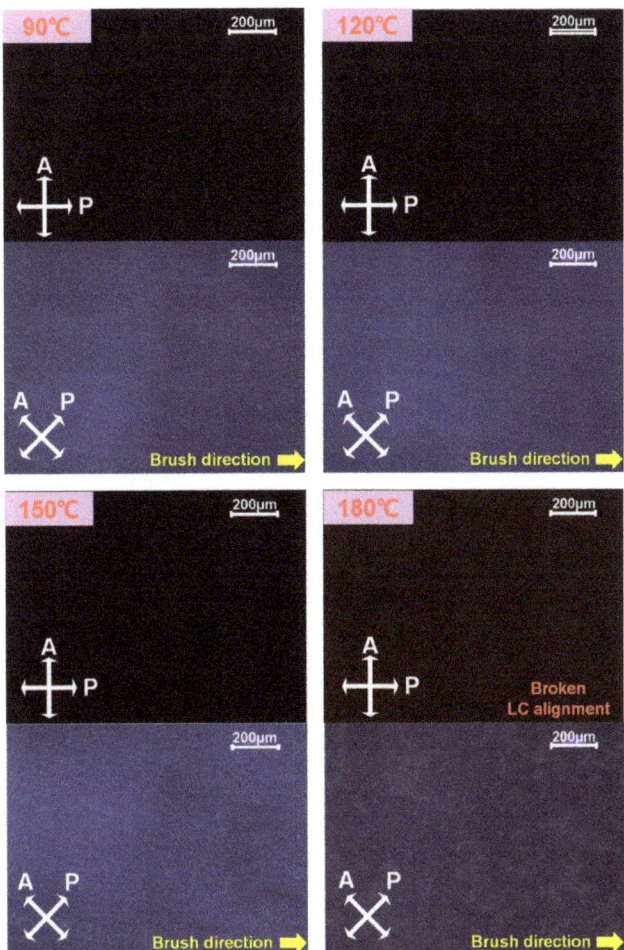

Figure 8. Thermal endurance of the brush-coated nickel yttrium oxide film to uniform LC alignment. Annealing was conducted from 90 to 180 °C at intervals of 30 °C for 10 min at each temperature.

The EO characteristics of the brush-coated NYO film were investigated using the R-T and V-T graphs, which were obtained from a TN-LC cell made from the NYO alignment layer. In the TN-LC system, the LCs in the cell converted their state between fall and rise depending on the applied external voltage. Moreover, the light transmittance could be controlled from this state transition. The R-T graph (Figure 9a) revealed the stable switching performance of the NYO film-based TN-LC cell. The LC fall and rise state transition times were obtained to 13.7 and 3.6 ms in each, and the response time which obtained by the summation of rise and fall transition times was acquired as 17.3 ms. Moreover, the threshold voltage corresponding to 90% of optical transmittance was obtained as 0.9 V (Figure 9b). This value represents the enhanced voltage operating characteristic of the NYO film compared to that of conventional PI layers [31]. This also demonstrates the superior threshold voltage property compared to other recent studies of LC alignment layers [32,33]. Importantly, this low operating voltage results in low consumption of power. These results verified that the NYO film formed by brush-coating technique has a high potential for applications in TN-LC systems.

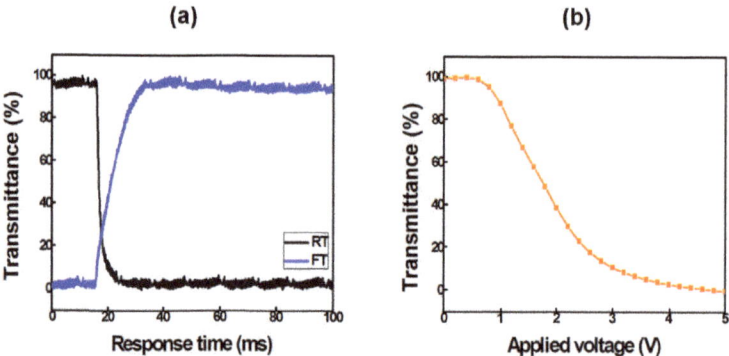

Figure 9. Investigation into electro-optical properties of twisted-nematic LC cell assembled from brush-coated nickel yttrium oxide films. (**a**) Response time and (**b**) threshold voltage.

4. Conclusions

An anisotropic NYO film surface was achieved using the brush coating method and was utilized as an LC alignment layer in this study. The NYO film was cured at 70, 150, and 230 °C after solution-processed brush coating, and the morphology of the formed films surface was investigated by AFM. The 230 °C cured film had a directional surface, and the XPS confirmed the well-formed NYO layer on the substrate. The directional surface was attributed to the shear stress, which was generated through the retracing force on the deposited solution. This nano/micro-grooved surface induced surface anisotropy, deriving uniform LC orientation on that. The homogeneous and uniform LC orientation state was demonstrated by POM and pre-tilt angle measurements. The NYO film achieved high optical transparency, and the contact angle analyses revealed an increase in hydrophilicity at higher curing temperatures. The LC cell fabricated by the NYO film exhibited suitable thermal endurance to LC alignment. Moreover, the NYO alignment layer also achieved good switching properties and enhanced operating voltage characteristics in a TN-LC system. Given these advantages, brush-coated NYO film has a high possibility for functional LC alignment layers.

Author Contributions: Conceptualization, D.-W.L.; methodology, D.-B.Y.; validation, D.-H.K.; formal analysis, D.-W.L.; investigation, J.-Y.O.; resources, D.-S.S.; writing—original draft preparation, D.-W.L.; visualization, J.-Y.O.; supervision, Y.L.; project administration, D.-S.S. All authors have read and agreed to the published version of the manuscript.

Funding: This research was supported by the National Research Foundation of Korea (NRF) grant funded by the Korea government (MSIT) (No. 2022R1F1A106419211).

Institutional Review Board Statement: Not applicable.

Informed Consent Statement: Not applicable.

Data Availability Statement: The data presented in this study are available upon request from the corresponding author.

Conflicts of Interest: The authors declare no conflict of interest.

References

1. Yan, X.; Mont, F.W.; Poxson, D.J.; Schubert, M.F.; Kim, J.K.; Cho, J.; Schubert, E.F. Refractive-index-matched indium-tin-oxide electrodes for liquid crystal displays. *Jpn. J. Appl. Phys.* **2009**, *48*, 120203. [CrossRef]
2. Li, J.; Wen, C.-H.; Gauza, S.; Lu, R.; Wu, S.-T. Refractive indices of liquid crystals for display applications. *J. Disp. Technol.* **2005**, *1*, 51. [CrossRef]
3. Lin, T.-H.; Jau, H.-C. Electrically controllable laser based on cholesteric liquid crystal with negative dielectric anisotropy. *Appl. Phys. Lett.* **2006**, *88*, 061122. [CrossRef]
4. Chen, Y.; Xu, D.; Wu, S.-T.; Yamamoto, S.; Haseba, Y.A. A low voltage and submillisecond-response polymer-stabilized blue phase liquid crystal. *Appl. Phys. Lett.* **2013**, *102*, 141116. [CrossRef]
5. Lee, J.J.; Park, H.G.; Han, J.J.; Kim, D.H.; Seo, D.-S. Surface reformation on solution-derived zinc oxide films for liquid crystal systems via ion-beam irradiation. *J. Mater. Chem. C* **2013**, *1*, 6824–6828. [CrossRef]
6. Garbovskiy, Y. Switching between purification and contamination regimes governed by the ionic purity of nanoparticles dispersed in liquid crystals. *Appl. Phys. Lett.* **2016**, *108*, 121104. [CrossRef]
7. Lee, W.-K.; Choi, Y.S.; Kang, Y.-G.; Sung, J.; Seo, D.-S.; Park, C. Super-fast switching of twisted nematic liquid crystals on 2D single wall carbon nanotube networks. *Adv. Funct. Mater.* **2011**, *21*, 3843–3850. [CrossRef]
8. Oh, S.-W.; Yoon, T.-H. Elimination of light leakage over the entire viewing cone in a homogeneously-aligned liquid crystal cell. *Opt. Express* **2014**, *22*, 5808. [CrossRef]
9. Outram, B.I.; Elston, S.J. Spontaneous and stable uniform lying helix liquid-crystal alignment. *J. Appl. Phys.* **2013**, *113*, 043103. [CrossRef]
10. Varghese, S.; Narayanankutty, S.; Bastiaansen, C.W.M.; Crawford, G.P.; Broer, D.J. Patterned Alignment of Liquid Crystals by μ-rubbing. *Adv. Mater.* **2004**, *16*, 1600. [CrossRef]
11. Kim, Y.J.; Zhuang, Z.; Patel, J.S. Effect of multidirection rubbing on the alignment of nematic liquid crystal. *Appl. Phys. Lett.* **2000**, *77*, 513. [CrossRef]
12. Ho, J.Y.L.; Chigrinov, V.G.; Kwok, H.S. Variable liquid crystal pretilt angles generated by photoalignment of a mixed polyimide alignment layer. *Appl. Phys. Lett.* **2007**, *90*, 243506. [CrossRef]
13. Ouchi, Y.; Lee, J.; Takezoe, H.; Fukuda, A.; Kondo, K.; Kitamura, T.; Mukoh, A. Smectic Layer Structure of Thin Ferroelectric Liquid Crystal Cells Aligned by SiO Oblique Evaporation Technique. *Jpn. J. Appl. Phys.* **1998**, *27*, L1993. [CrossRef]
14. Yaroshchuk, O.; Kravchuk, R.; Dobrovolskyy, A.; Qiu, L.; Lavrentovich, O.D. Planar and tilted uniform alignment of liquid crystals by plasma-treated substrates. *Liq. Cryst.* **2004**, *31*, 859. [CrossRef]
15. Haaren, J.V. Wiping out dirty displays. *Nature* **2001**, *411*, 29–30. [CrossRef] [PubMed]
16. Rao, K.V.; Smakula, A. Dielectric properties of cobalt oxide, nickel oxide, and their mixed crystals. *J. Appl. Phys.* **1965**, *36*, 2031. [CrossRef]
17. Tsutsumi, T. Dielectric properties of Y_2O_3 thin films prepared by vacuum evaporation. *Jpn. J. Appl. Phys.* **1970**, *9*, 735. [CrossRef]
18. Mell, C.C.; Finn, S.R. Forces exerted during the brushing of a paint. *Rheol. Acta* **1965**, *4*, 260. [CrossRef]
19. Kim, S.-S.; Na, S.-I.; Jo, J.; Tae, G.; Kim, D.-Y. Efficient polymer solar cells fabricated by simple brush painting. *Adv. Mater.* **2007**, *19*, 4410. [CrossRef]
20. Lee, D.W.; Kim, D.H.; Oh, J.Y.; Kim, D.-H.; Liu, Y.; Seo, D.-S. Tunable liquid crystal alignment and driving mode on lanthanum aluminum zirconium zinc-oxide thin film achieved by convenient brush-coating method. *ChemNanoMat* **2022**, *8*, e202200131. [CrossRef]
21. Lee, D.W.; Kim, E.M.; Heo, G.S.; Kim, D.H.; Oh, J.Y.; Kim, D.-H.; Liu, Y.; Seo, D.-S. Oriented Yttrium Strontium Tin Oxide Micro/Nanostructures Induced by Brush-Coating for Low-Voltage Liquid Crystal Systems. *ACS Appl. Nano Mater.* **2022**, *5*, 6925. [CrossRef]
22. Han, K.Y.; Miyashita, T.; Uchida, T. Accurate measurement of the pretilt angle in a liquid crystal cell by an improved crystal rotation method. *Mol. Cryst. Liq. Cryst. Sci. Technol. Sect. A Mol. Cryst. Liq. Cryst.* **1994**, *241*, 147–157. [CrossRef]
23. Chen, K.-H.; Chang, W.-Y.; Chen, J.-H. Measurement of the pretilt angle and the cell gap of nematic liquid crystal cells by heterodyne interferometry. *Opt. Express* **2009**, *17*, 14143. [CrossRef] [PubMed]
24. Fukuda, J.-I.; Yoneya, M.; Yokoyama, H. Surface-Groove-Induced Azimuthal Anchoring of a Nematic Liquid Crystal: Berreman's Model Reexamined. *Phys. Rev. Lett.* **2007**, *98*, 187803. [CrossRef] [PubMed]
25. Kikuchi, H.; Logan, J.A.; Yoon, D.Y. Study of local stress, morphology, and liquid-crystal alignment on buffed polyimide surfaces. *J. Appl. Phys.* **1996**, *79*, 6811. [CrossRef]
26. Berreman, D.W. Solid Surface Shape and the Alignment of an Adjacent Nematic Liquid Crystal. *Phys. Rev. Lett.* **1972**, *28*, 1683. [CrossRef]
27. Chae, B.; Kim, S.B.; Lee, S.W.; Kim, S.I.; Choi, W.; Lee, B.; Ree, M.; Lee, K.H.; Jung, J.C. Surface morphology, molecular reorientation, and liquid crystal alignment properties of rubbed nanofilms of a well-defined brush polyimide with a fully rodlike backbone. *Macromolecules* **2002**, *35*, 10119. [CrossRef]

28. Adamson, A.W. *Physical Chemistry of Surfaces*, 5th ed.; Wiley-Interscience: Hoboken, NJ, USA, 1990.
29. Li, J.; Pan, Y.; Xiang, C.; Ge, Q.; Guo, J. Low temperature synthesis of ultrafine α-Al_2O_3 powder by a simple aqueous sol–gel process. *Ceram. Int.* **2006**, *32*, 587. [CrossRef]
30. Mun, H.-Y.; Jeong, H.-C.; Lee, J.H.; Won, J.-H.; Park, H.-G.; Oh, B.-Y.; Seo, D.-S. Poly(styrene–maleic anhydride) films as alignment layers for liquid crystal systems via ion-beam irradiation. *RSC Adv.* **2016**, *6*, 76743. [CrossRef]
31. Jeong, H.-C.; Lee, J.H.; Won, J.; Oh, B.Y.; Kim, D.H.; Lee, D.W.; Song, I.H.; Liu, Y.; Seo, D.-S. One-Dimensional Surface Wrinkling for Twisted Nematic Liquid Crystal Display Based on Ultraviolet Nanoimprint Lithography. *Opt. Express* **2019**, *27*, 18096. [CrossRef]
32. Wang, Y.-F.; Guo, Y.-Q.; Ren, Y.-X.; Fu, M.-Z.; Zhu, J.-L.; Sun, Y.-B. Study on polyvinylidene fluoride as alignment layer in twist-nematic liquid crystal display. *Liq. Cryst.* **2018**, *45*, 857. [CrossRef]
33. John V, N.; Rajeev, S.P.; Varghese, S. Ferroelectric polymer nanocomposite alignment layer in twisted nematic liquid crystal devices for reducing switching voltage. *Liq. Cryst.* **2019**, *46*, 736. [CrossRef]

Article

Synthesis, Mesomorphic Properties and Application of (R,S)-1-Methylpentyl 4′-Hydroxybiphenyl-4-carboxylate Derivatives

Magdalena Urbańska * and Mateusz Szala

Institute of Chemistry, Military University of Technology, 00-908 Warsaw, Poland
* Correspondence: magdalena.urbanska@wat.edu.pl

Abstract: Thirteen new liquid crystalline racemic mixtures were synthesized and investigated. For these racemic mixtures, the phase sequences and their changes were determined by polarizing optical microscopy (POM). The phase transition temperatures and transition enthalpies were checked by differential scanning calorimetry (DSC). All new racemates have an anticlinic smectic C_A phase in a broad temperature range. Three highly tilted antiferroelectric mixtures were doped with six racemates at a concentration of 20% by weight. The helical pitch of the prepared mixtures was measured by the spectrophotometry method. All doped mixtures have a longer helical pitch than the base mixtures.

Keywords: synthesis; racemates; anticlinic phase; antiferroelectric mixtures; helical pitch

Citation: Urbańska, M.; Szala, M. Synthesis, Mesomorphic Properties and Application of (R,S)-1-Methylpentyl 4′-Hydroxybiphenyl-4-carboxylate Derivatives. *Crystals* **2022**, *12*, 1710. https://doi.org/10.3390/cryst12121710

Academic Editors: Zhenghong He and Yuriy Garbovskiy

Received: 11 October 2022
Accepted: 23 November 2022
Published: 25 November 2022

Publisher's Note: MDPI stays neutral with regard to jurisdictional claims in published maps and institutional affiliations.

Copyright: © 2022 by the authors. Licensee MDPI, Basel, Switzerland. This article is an open access article distributed under the terms and conditions of the Creative Commons Attribution (CC BY) license (https:// creativecommons.org/licenses/by/ 4.0/).

1. Introduction

The racemic mixture, also called racemate, is the mixture of equal quantities of two enantiomers or substances that have dissymmetric molecular structures, i.e., mirror images of one another (Figure 1). Each enantiomer rotates the plane of polarization of plane-polarized light through a characteristic angle, but because the rotatory effect of each component exactly cancels that of the other, the racemic mixture is optically inactive [1,2].

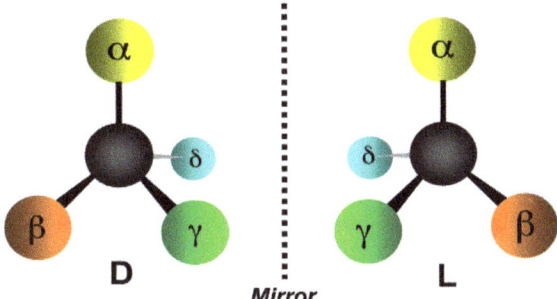

Figure 1. Example of D- and L-enantiomers.

The racemates can be used as dopants for the liquid crystalline mixtures, improving some of their properties, such as the antiferroelectric phase range, the helical pitch length, the reduction of the rotational viscosity, the decrease of the values of the spontaneous polarization, etc. [3–5]. Liquid crystalline mixtures are widely used in optical devices; therefore, new materials are constantly being developed for this purpose [6–11]. The use of the racemates also leads to a reduction in the preparation costs of the liquid crystalline mixtures. They can also be separated by chiral chromatography (HPLC, UPLC, SFC) [12–19].

Herein, thirteen new racemic mixtures with the acronym: n.(X_1X_2) (R,S) and the general formula shown in Figure 2a are synthesized, and their mesomorphic and thermodynamic properties are measured and discussed.

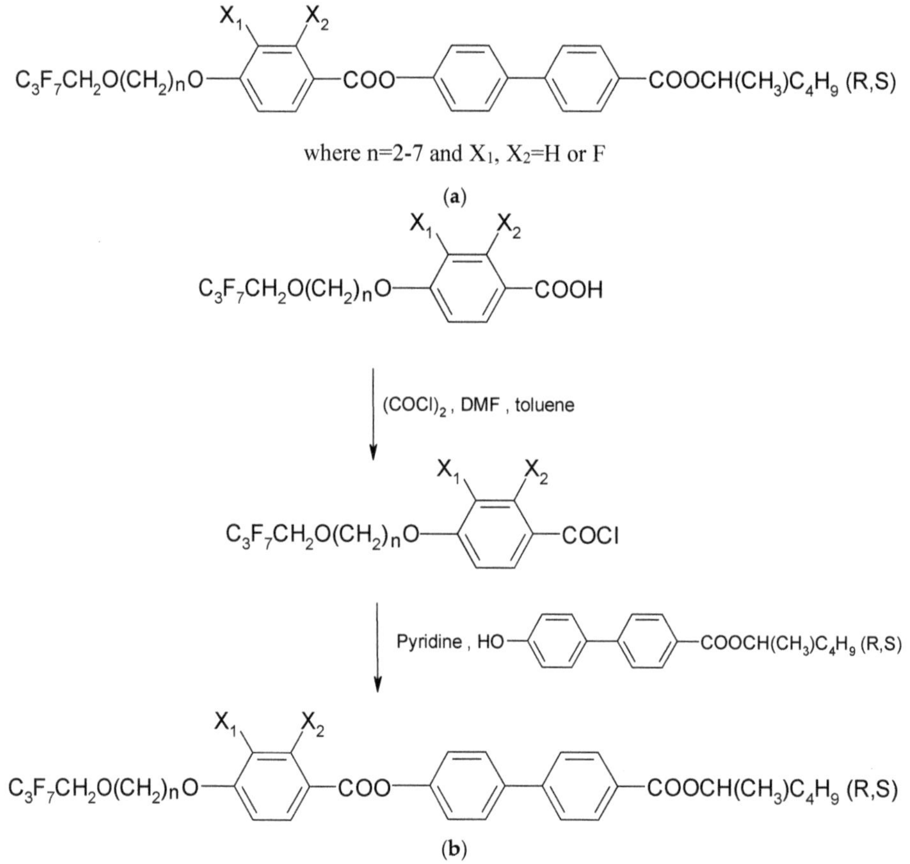

Figure 2. (a) The general formula of the synthesized racemic mixtures. (b) Scheme of the synthesis of the new racemic mixtures.

Six of the racemates are used as dopants for three highly tilted antiferroelectric liquid crystalline mixtures. The base mixtures are described in Refs. [20,21]. All the components of the base mixtures are (S) enantiomers belonging to the same homologous series. The aim of the work is to discuss the mesomorphic properties of the racemates with a wide-temperature anticlinic phase and to demonstrate the advisability of using the racemates as dopants for the antiferroelectric mixtures. If so, it would be possible to extend the range of the antiferroelectric phase and thus increase the helical pitch, which is quite sensitive to doping [3,4,22–25]. An analysis of the obtained results is presented.

2. Synthesis of the Racemic Mixtures

In this work, thirteen previously unpublished racemates were synthesized by treating (R,S) 4′-(1-methylpentyloxycarbonyl)biphenol with benzoic acid chloride, see Figure 2b. The synthesis of the racemates was carried out as described in Ref. [3]. The benzoic acids were synthesized using the method described in Ref. [26].

For the synthesis of (R,S) 4′-(1-methylpentyloxycarbonyl)biphenol, the method described in Ref. [27] was chosen. The commercially available (R,S)-2-hexanol was used for the synthesis.

The purity of the synthesized racemates was checked using a Shimadzu prominence chromatograph with an SPD-M20A diode array detector. The purity of the racemates was also monitored by thin-layer chromatography (silica gel on aluminum).

The purity of the racemates, determined by HPLC, is shown in Table 1. About 1 g of the final racemates were synthesized in all cases. The purity is 99% or more.

Table 1. The purity of the synthesized racemic mixtures.

The Acronym of the Racemic Mixture	Purity [%]
2.(HH) (R,S)	99.0
2.(HF) (R,S)	99.0
2.(FH) (R,S)	99.7
3.(HF) (R,S)	99.5
3.(FF) (R,S)	99.5
4.(HH) (R,S)	99.9
4.(HF) (R,S)	99.0
5.(HH) (R,S)	99.8
5.(HF) (R,S)	99.8
5.(FF) (R,S)	99.8
6.(FH) (R,S)	99.3
6.(FF) (R,S)	99.7
7.(HH) (R,S)	99.2

The structure of the racemic mixtures was confirmed by ^{1}H NMR and ^{13}C NMR nuclear magnetic resonance. The NMR spectra were acquired on a Bruker Avance III 500 MHz spectrometer. This device has a superconducting magnet, which generates a magnetic field at induction 11.75 T, and for the sample radiation effects at a frequency of 500 MHz for protons and 125 MHz for carbon nuclei. The deuterated chloroform (CDCl$_3$) as the solvent was used. The spectra of all of the samples were measured at 25 °C.

The ^{1}H NMR and ^{13}C NMR spectra of all racemates were added to the Supplementary Materials (Figures S1–S26). Details of the basic chemical characterization of the new racemates are presented below (the acronyms of the racemates are used).

2.(HH) (R,S)
^{1}H NMR [ppm]: 0.94 (t, -CH$_3$), 1.39 (d, -CH$_2$-); 1.66 (m, -CH$_2$-); 1.78 (m, -CH$_2$-); 4.06 (t, -CH$_2$-); 4.13 (t, -CH$_2$-); 4.06-4.27 (t, -CH$_2$-); 5.21 (sext, -CH-), 7.02 (d, C$_{Ar}$H), 7.34 (d, C$_{Ar}$H), 7.68 (q, C$_{Ar}$H), 8.15 (dd, C$_{Ar}$H).
^{13}C {^{1}H} NMR [ppm]: 14.0 (s, CH$_3$); 20.1 (s, CH$_3$); 22.6 (s, CH$_2$), 27.6 (s, CH$_2$), 35.8 (s, CH$_2$), 67.6 (s, CH$_2$), 68.2 (t, CF), 71.2 (s, CH$_2$), 71.8 (s, CH), 114.4 (s, C$_{Ar}$H), 122.2 (s, C$_{Ar}$H), 122.3 (s, C$_{Ar}$), 126.9 (s, C$_{Ar}$H), 128.3 (s, C$_{Ar}$H), 129.8 (s, C$_{Ar}$), 130.1 (s, C$_{Ar}$H), 132.4 (s, C$_{Ar}$H), 137.8 (s, C$_{Ar}$), 144.6 (s, C$_{Ar}$), 151.1 (s, C$_{Ar}$), 162.9 (s, C$_{Ar}$), 164.8 (s, C$_{Ar}$), 166.1 (s, C$_{Ar}$).

2.(HF) (R,S)
^{1}H NMR [ppm]: 0.94 (t, -CH$_3$), 1.39 (d, -CH$_2$-); 1.61 (m, -CH$_2$-); 1.78 (m, -CH$_2$-); 4.06 (t, -CH$_2$-); 4.12 (t, -CF-); 4.25 (t, -CH$_2$-); 5.22 (sext, -CH-), 6.77 (dd, C$_{Ar}$H), 7.35 (d, C$_{Ar}$H), 7.68 (q, C$_{Ar}$H), 8.09 (s, C$_{Ar}$H), 8.11 (s, C$_{Ar}$H), 8.13 (s, C$_{Ar}$H), 8.14 (s, C$_{Ar}$H).
^{13}C {^{1}H} NMR [ppm]: 14.0 (s, CH$_3$); 20.1 (s, CH$_3$); 22.6 (s, CH$_2$), 27.6 (s, CH$_2$), 35.8 (s, CH$_2$), 67.9 (s, CH$_2$), 68.2 (t, CF), 71.0 (s, CH), 71.8 (s, CH), 103.1 (d, C$_{Ar}$F), 110,6 (d, C$_{Ar}$F), 110.8 (d, C$_{Ar}$F), 122.2 (s, C$_{Ar}$H), 126.9 (s, C$_{Ar}$H), 128.3 (s, C$_{Ar}$H), 129.8 (s, C$_{Ar}$), 130.1 (s, C$_{Ar}$H), 134.0

(s, C$_{Ar}$), 137.9 (s, C$_{Ar}$), 144.6 (s, C$_{Ar}$), 150.8 (s, C$_{Ar}$), 162.3 (s, C$_{Ar}$), 162.4 (s, C$_{Ar}$), 164.0 (d, CF), 164.9 (s, C$_{Ar}$), 166.1 (s, C$_{Ar}$).

2.(FH) (R,S)
^{1}H NMR [ppm]: 0.94 (t, -CH$_3$); 1.38 (d, -CH$_2$-); 1.66 (m, -CH$_2$-); 1.78 (m, -CH$_2$-), 4.09 (t, -CH$_2$-), 4.16 (t, -CH$_2$-), 4.35 (t, -CH$_2$-), 5.22 (sext, -CH-); 7.09 (t, C$_{Ar}$H), 7.32 (d, C$_{Ar}$H), 7.69 (t, C$_{Ar}$H), 7.95–8.02 (dd, C$_{Ar}$H); 8.13 (d, C$_{Ar}$H).
^{13}C {^{1}H} NMR [ppm]: 13.9 (s, CH$_3$); 20.1 (s, CH$_3$), 22.6–35.8 (s, CH$_2$), 68.4 (t, CF); 69.0 (s, CH); 71.1 (s, CH); 71.8 (s, CH); 122.2 (s, C$_{Ar}$H), 122.7 (d, C$_{Ar}$F), 126.9 (s, C$_{Ar}$H), 127.4 (d, C$_{Ar}$F), 150.9 (d, C$_{Ar}$F), 151.2 (d, C$_{Ar}$F), 152.9 (s, C$_{Ar}$), 163.0 (d, C$_{Ar}$F), 166.0 (s, C$_{Ar}$).

3.(HF) (R,S)
^{1}H NMR [ppm]: 0.94 (t, -CH$_3$); 1.38 (d, -CH$_2$-); 1.65 (m, -CH$_2$-); 1.78 (m, -CH$_2$-); 2.15 (m, -CH$_2$-); 3.83 (t, -CH$_2$-); 3.98 (t, -CH$_2$-); 4.17 (t, -CH$_2$-); 5.21 (sext, -CH-); 6.7–6.9 (dd, C$_{Ar}$H); 7.34 (d, C$_{Ar}$H), 7.68 (q, C$_{Ar}$H), 8.09 (t, C$_{Ar}$H), 8.13 (d, C$_{Ar}$H).
^{13}C {^{1}H} NMR [ppm]: 14.0 (s, CH$_3$); 20.1 (s, CH$_3$), 22.6–35.8 (s, CH$_2$); 64.9 (s, CH$_2$); 67.8 (t, CF), 69.1 (s, CH); 71.8 (s, CH); 102.9 (d, CF), 110.1 (d, C$_{Ar}$F); 110.7 (d, C$_{Ar}$F); 162.4 (d, C$_{Ar}$F); 162.9 (s, C$_{Ar}$), 164.5 (d, C$_{Ar}$F), 165.0 (s, C$_{Ar}$), 166.1 (s, C$_{Ar}$).

3.(FF) (R,S)
^{1}H NMR [ppm]: 0.94 (t, -CH$_3$), 1.39 (d, -CH$_2$-); 1.66 (m, -CH$_2$-); 1.78 (m, -CH$_2$-); 2.19 (sext, -CH-), 3.86 (t, -CH$_2$-), 3.99 (t, -CH$_2$-), 4.28 (t, -CH$_2$-), 5.21 (sext, -CH-), 6.88 (t, C$_{Ar}$H); 7.34 (d, C$_{Ar}$H); 7.69 (t, C$_{Ar}$H); 7.90 (t, C$_{Ar}$H); 8.13 (d, C$_{Ar}$H).
^{13}C {^{1}H} NMR [ppm]: 13.9 (s, CH$_3$); 20.1 (s, CH$_3$); 22.6–35.8 (s, CH$_2$), 66.0 (s, CH$_2$), 67.8 (t, CH), 68.8 (s, CH$_2$), 71.8 (s, CH), 108.5 (s, C$_{Ar}$H), 122.2 (s, C$_{Ar}$H), 126.9 (s, C$_{Ar}$H), 127.1 (d, C$_{Ar}$F), 128.4 (s, C$_{Ar}$H), 129.9 (s, C$_{Ar}$H), 130.1 (s, C$_{Ar}$H), 138.1 (s, C$_{Ar}$H), 140.4 (dd, C$_{Ar}$F), 144.5 (s, C$_{Ar}$), 150.6 (s, C$_{Ar}$), 150.8 (d, C$_{Ar}$F), 161.9 (s, C$_{Ar}$), 166.0 (s, C$_{Ar}$).

4.(HH) (R,S)
^{1}H NMR [ppm]: 0.94 (t, -CH$_3$), 1.39 (d, -CH$_2$-); 1.66 (m, -CH$_2$-); 1.78 (m, -CH$_2$-); 1.85 (m, -CH$_2$-); 1.96 (m, -CH$_2$-); 3.72 (t, -CH$_2$-); 3.97 (t, -CH$_2$-); 4.11 (t, -CH$_2$-); 5.21 (sext, -CH-), 7.00 (d, C$_{Ar}$H), 7.33 (d, C$_{Ar}$H), 7.69 (q, C$_{Ar}$H), 8.13 (d, C$_{Ar}$H), 8.18 (d, C$_{Ar}$H).
^{13}C {^{1}H} NMR [ppm]: 14.0 (s, CH$_3$); 20.1 (s, CH$_3$); 22.6 (s, CH$_2$), 25.7 (s, CH$_2$), 26.2 (s, CH$_2$), 27.6 (s, CH$_2$), 35.8 (s, CH$_2$), 67.8 (s, CH$_2$), 71.8 (s, CH), 72.6 (s, CH), 114.3 (s, C$_{Ar}$H), 121.6 (s, C$_{Ar}$), 122.3 (s, C$_{Ar}$H), 126.9 (s, C$_{Ar}$H), 128.3 (s, C$_{Ar}$H), 129.7 (s, C$_{Ar}$), 130.1 (s, C$_{Ar}$H), 132.4 (s, C$_{Ar}$H), 137.6 (s, C$_{Ar}$), 144.6 (s, C$_{Ar}$), 151.2 (s, C$_{Ar}$), 163.5 (s, C$_{Ar}$), 164.9 (s, C$_{Ar}$), 166.1 (s, C$_{Ar}$).

4.(HF) (R,S)
^{1}H NMR [ppm]: 0.94 (t, -CH$_3$), 1.39 (d, -CH$_2$-); 1.66 (m, -CH$_2$-); 1.82 (m, -CH$_2$-), 1.96 1.82 (m, -CH$_2$-), 3.71 1.82 (t, -CH$_2$-), 3.96 1.82 (t, -CH$_2$-), 4.08 1.82 (m, -CH$_2$-), 5.20 (sext, -CH-), 6.70 (dd, C$_{Ar}$H), 6.81 (dd, C$_{Ar}$H), 7.33 (d, C$_{Ar}$H), 7.67 (q, C$_{Ar}$H), 8.08 (t, C$_{Ar}$H), 8.13 (d, C$_{Ar}$H).
^{13}C {^{1}H} NMR [ppm]: 13.9 (s, CH$_3$); 20.0 (s, CH$_3$); 22.6 (s, CH$_2$); 25.6 (s, CH$_2$); 26.1 (s, CH$_2$); 27.6 (s, CH$_2$); 35.8 (s, CH$_2$); 67.6 (t, CF); 68.3 (s, CH$_2$); 71.8 (s, CH); 72.6 (s, CH); 102.8 (d, CF); 109.8 (d, CF); 110.8 (d, CF); 122.3 (s, C$_{Ar}$H), 126.9 (s, C$_{Ar}$H), 128.3 (s, C$_{Ar}$H), 129.7 (s, C$_{Ar}$), 130.1 (s, C$_{Ar}$H), 133.9 (s, C$_{Ar}$), 137.8 (s, C$_{Ar}$), 144.6 (s, C$_{Ar}$), 150.8 (s, C$_{Ar}$), 162.4 (d, C$_{Ar}$F), 162.9 (s, C$_{Ar}$), 164.7 (d, C$_{Ar}$F), 165.0 (s, C$_{Ar}$), 166.1 (s, C$_{Ar}$).

5.(HH) (R,S)
^{1}H NMR [ppm]: 0.94 (t, -CH$_3$), 1.39 (d, -CH$_2$-); 1.63 (m, -CH$_2$-); 1.74 (m, -CH$_2$-); 1.89 (sext, -CH-), 3.67 (t, -CH$_2$-); 3.95 (t, -CH$_2$-); 4.09 (t, -CH$_2$-); 5.20 (sext, -CH-), 7.00 (d, C$_{Ar}$H), 7.32 (d, C$_{Ar}$H), 7.68 (q, C$_{Ar}$H), 8.13 (d, C$_{Ar}$H), 8.18 (d, C$_{Ar}$H).
^{13}C {^{1}H} NMR [ppm]: 14.0 (s, CH$_3$); 20.1 (s, CH$_3$), 22.4 (s, CH$_2$), 22.6 (s, CH$_2$), 27.6 (s, CH$_2$), 28.8 (s, CH$_2$), 29.2 (s, CH$_2$), 35.8 (s, CH$_2$), 67.3 (t, CF), 68.0 (s, CH$_2$), 71.8 (s, CH), 72.9 (s, CH), 114.3 (s, C$_{Ar}$H), 121.5 (s, C$_{Ar}$), 122.3 (s, C$_{Ar}$H), 126.9 (s, C$_{Ar}$H), 128.3 (s, C$_{Ar}$H), 129.7 (s, C$_{Ar}$), 130.1 (s, C$_{Ar}$H), 132.4 (s, C$_{Ar}$H), 137.7 (s, C$_{Ar}$), 144.6 (s, C$_{Ar}$), 151.2 (s, C$_{Ar}$), 163.5 (s, C$_{Ar}$), 164.9 (s, C$_{Ar}$), 166.1 (s, C$_{Ar}$).

5.(FH) (R,S)
^{1}H NMR [ppm]: 0.94 (t, -CH$_3$), 1.39 (d, -CH$_2$-); 1.64 (m, -CH$_2$-); 1.74 (m, -CH$_2$-); 1.93 (m, -CH$_2$-); 3.67 (t, -CH$_2$-); 3.95 (t, -CF-); 4.16 (t, -CH$_2$-); 5.21 (sext, -CH-), 7.06 (t, C$_{Ar}$H), 7.31 (d, C$_{Ar}$H), 7.68 (t, C$_{Ar}$H), 7.93 (dd, C$_{Ar}$H), 8.00 (d, C$_{Ar}$H), 8.13 (d, C$_{Ar}$H).
^{13}C {^{1}H} NMR [ppm]: 14.0 (s, CH$_3$), 20.1 (s, CH$_3$); 22.3 (s, CH$_2$), 22.6 (s, CH$_2$), 27.6 (s, CH$_2$), 28.7 (s, CH$_2$), 29.2 (s, CH$_2$), 35.8 (s, CH$_2$), 67.6 (t, CF), 69.2 (s, CH$_2$), 71.8 (s, CH), 72.9 (s, CH), 113.4 (s, C$_{Ar}$), 117.8 (d, C$_{Ar}$F), 121.8 (s, C$_{Ar}$F), 121.9 (s, C$_{Ar}$H), 126.9 (s, C$_{Ar}$H), 127.4 (d, C$_{Ar}$F), 128.4 (s, C$_{Ar}$H), 129.8 (s, C$_{Ar}$), 130.1 (s, C$_{Ar}$H), 137.9 (s, C$_{Ar}$), 144.5 (s, C$_{Ar}$), 150.9 (s, C$_{Ar}$H), 151.8 (d, C$_{Ar}$F), 152.9 (s, C$_{Ar}$), 164.0 (d, C$_{Ar}$), 166.0 (s, C$_{Ar}$).

5.(FF) (R,S)
^{1}H NMR [ppm]: 0.94 (t, -CH$_3$), 1.39 (d, -CH$_2$-); 1.62 (m, -CH$_2$-); 1.75 (m, -CH$_2$-); 1.94 (m, -CH$_2$-); 3.67 (t, -CH$_2$-); 3.83 (t, -CH$_2$-); 4.18 (t, -CH$_2$-), 5.21 (sext, -CH-); 6.85 (t, C$_{Ar}$H); 7.36 (d, C$_{Ar}$H), 7.69 (t, C$_{Ar}$H); 7.89 (t, C$_{Ar}$H), 8.13 (d, C$_{Ar}$H).
^{13}C {^{1}H} NMR [ppm]: 14.0 (s, CH$_3$), 20.1 (s, CH$_3$), 22.3–35.8 (s, CH$_2$); 67.6 (t, CF), 69.7 (s, CH); 71.8 (s, CH); 72.9 (s, CH), 108.4 (s, CH), 111.3 (d, C$_{Ar}$F), 122.2 (s, C$_{Ar}$H), 126.9 (s, C$_{Ar}$H), 127.0 (d, C$_{Ar}$F), 129.9 (s, C$_{Ar}$H), 130.1 (s, C$_{Ar}$H), 138.0 (s, C$_{Ar}$), 144.5 (s, C$_{Ar}$), 150.6 (s, C$_{Ar}$).

6.(FH) (R,S)
^{1}H NMR [ppm]: 0.94 (t, -CH$_3$), 1.39 (d, -CH$_2$-); 1.50 (m, -CH$_2$-); 1.56 (m, -CH$_2$-), 1.67 (m, -CH$_2$-); 1.78 (m, -CH$_2$-), 1.91 (m, -CH$_2$-); 3.64 (t, -CH$_2$-); 3.94 (t, -CH$_2$-); 4.15 (t, -CH$_2$-); 5.20 (sext, -CH-), 7.06 (t, C$_{Ar}$H), 7.31 (d, C$_{Ar}$H), 7.69 (q, C$_{Ar}$H), 7.93 (dd, C$_{Ar}$H), 7.99 (d, C$_{Ar}$H), 8.13 (d, C$_{Ar}$H).
^{13}C {^{1}H} NMR [ppm]: 13.9 (s, CH$_3$), 20.1 (s, CH$_3$); 22.6 (s, CH$_2$); 25.5 (d, CF); 27.6 (s, CH$_2$); 28.9 (s, CH$_2$); 29.3 (s, CH$_2$), 35.8 (s, CH$_2$), 67.6 (t, CF), 69.2 (s, CH$_2$), 71.8 (s, CH), 73.0 (s, CH), 113.5 (s, C$_{Ar}$), 117.7 (d, C$_{Ar}$F), 121.7 (s, C$_{Ar}$F), 122.2 (s, C$_{Ar}$H), 126.9 (s, C$_{Ar}$H), 127.4 (s, C$_{Ar}$), 128.4 (s, C$_{Ar}$H), 129.8 (s, C$_{Ar}$), 130.1 (s, C$_{Ar}$H), 137.9 (s, C$_{Ar}$), 144.5 (s, C$_{Ar}$), 150.9 (d, CF), 151.9 (d, CF), 152.9 (s, C$_{Ar}$), 164.0 (d, CF), 166.0 (s, C$_{Ar}$).

6.(FF) (R,S)
^{1}H NMR [ppm]: 0.94 (t, -CH$_3$), 1.39 (d, -CH$_2$-); 1.48 (m, -CH$_2$-); 1.55 (m, -CH$_2$-), 1.66 (m, -CH$_2$-), 1.78 (m, -CH$_2$-), 1.91 (quin, -CH$_2$-); 3.64 (t, -CH$_2$-); 3.94 (t, -CH$_2$-); 4.16 (t, -CH$_2$-); 5.20 (sext, -CH-), 6.86 (t, C$_{Ar}$H), 7.3 (d, C$_{Ar}$H), 7.69 (t, C$_{Ar}$H), 7.88 (t, C$_{Ar}$H), 8.13 (d, C$_{Ar}$H).
^{13}C {^{1}H} NMR [ppm]: 13.9 (s, CH$_3$), 20.1 (s, CH$_3$); 22.6 (s, CH$_2$); 25.5 (d, CF); 27.6 (s, CH$_2$); 28.8 (s, CH$_2$); 29.3 (s, CH$_2$), 35.8 (s, CH$_2$), 67.6 (t, CF); 69.8 (s, CH$_2$), 71.8 (s, CH), 72.9 (s, CH), 108.5 (s, C), 111.2 (d, CF), 122.1 (s, C$_{Ar}$H), 126.9 (s, C$_{Ar}$H), 127.04 (s, C$_{Ar}$), 128.4 (s, C$_{Ar}$H), 129.9 (s, C$_{Ar}$), 130.1 (s, C$_{Ar}$H), 138.0 (s, C$_{Ar}$), 140.5 (d, CF), 142.5 (d, C$_{Ar}$F), 144.5 (s, C$_{Ar}$), 150.7 (s, C$_{Ar}$), 153.1 (s, C$_{Ar}$), 161.9 (s, C$_{Ar}$), 166.0 (s, C$_{Ar}$).

7.(HH) (R,S)
^{1}H NMR [ppm]: 0.94 (t, -CH$_3$), 1.39 (d, -CH$_2$-); 1.52 (m, -CH$_2$-); 1.66 (m, -CH$_2$-); 1.78 (m, -CH$_2$-); 1.85 (sext, -CH-), 3.62 (t, -CH$_2$-); 3.94 (t, -CH$_2$-); 4.07 (t, -CH$_2$-); 5.21 (sext, -CH-), 7.00 (d, C$_{Ar}$H), 7.32 (d, C$_{Ar}$H), 7.69 (q, C$_{Ar}$H), 1.18 (d, C$_{Ar}$H), 8.19 (d, C$_{Ar}$H).
^{13}C {^{1}H} NMR [ppm]: 14.0 (s, CH$_3$); 20.1 (s, CH$_3$); 22.6 (s, CH$_2$), 25.7 (s, CH$_2$), 25.9 (s, CH$_2$), 27.6 (s, CH$_2$), 29.0 (d, CF), 29.4 (s, CH$_2$), 35.8 (s, CH$_2$), 67.6 (t, CF), 68.2 (s, CF), 71.8 (s, CH), 73.1 (s, CH), 114.3 (s, C$_{Ar}$H), 121.4 114.4 (s, C$_{Ar}$), 122.3 (s, C$_{Ar}$H), 128.3 (s, C$_{Ar}$H), 129.8 (s, C$_{Ar}$), 130.1 (s, C$_{Ar}$H), 132.3 (s, C$_{Ar}$H), 137.7 (s, C$_{Ar}$), 144.6 (s, C$_{Ar}$), 151.2 (s, C$_{Ar}$), 163.6 (s, C$_{Ar}$), 164.9 (s, C$_{Ar}$), 166.0 (s, C$_{Ar}$).

3. Measurements

The mesomorphic properties of the racemates and the prepared mixtures were investigated using a polarizing optical microscopy technique. An OLYMPUS BX51 polarizing optical microscope equipped with a Linkam THMS-600 hot stage and a temperature controller TMS-93 were used. The phase sequences and the phase transition temperatures were determined by observation of the characteristic textures and their changes. In addition, differential scanning calorimetry (DSC) measurements were carried out using a Netzsch DSC 204 F1 Phoenix calorimeter. These measurements were performed in a nitrogen atmosphere under 2 °C·min^{-1} heating and cooling rate.

The helical pitch (p) value measurements were based on the phenomenon of wavelength-dependent selective reflection of light [28]. The selective reflection wavelength was determined in the temperature range of the tilted chiral smectic phases on samples placed on a glass plate coated with a surfactant-promoting homeotropic orientation of the molecular director on the substrate glass plate. The other surface of the smectic slab was left free, which also assures the homeotropic orientation at the air-LC contact boundary. The experimental details have been reported elsewhere [29,30]. The helical pitch was calculated according to Equations (1) and (2) for the antiferroelectric and ferroelectric phases, respectively.

$$\lambda_{max} = n \cdot p \qquad (1)$$

$$\lambda_{max} = 2n \cdot p \qquad (2)$$

Here λ_{max} is the wavelength at the center of the selective reflection waveband. The value of $n = 1.5$ was taken for the calculation [31,32]. The transmission spectra were acquired using a Shimadzu UV-VIS-NIR spectrometer UV-3600 at the wavelength range of 360–3000 nm. All the observations of the selective reflection spectra were carried out during the cooling cycle. An MLWU7 temperature controller with a Peltier element was used to drive the temperature within the range of 2–110 °C, with an accuracy of ±0.1 °C.

4. Mesomorphic Properties of the Racemic Mixtures

For the synthesized racemic mixtures, the phase sequences and the phase transition temperatures are summarized in Table 2 and visualized in Figure 3. The textures for the observed phases (the smectic A phase, the smectic C phase and the smectic C_A phase) are shown in Figure 4a–c.

Table 2. The phase transition temperatures [°C] and enthalpies [J·g^{-1}] of the synthesized racemic mixtures. ("*" means that the phase is observed and "-" means that it is not observed; The enthalpy is indicated by the italics).

Acronym	Cr		SmC$_A$		SmC		SmA		Iso
2.(HH) (R,S)	*	74.3 *4.9* *34.9*	*	138.6 140.3 *9.9*	-		-		*
2.(HF) (R,S)	*	54.5 - *30.1*	*	119.8 118.7 *9.1*	-		-		*
2.(FH) (R,S)	*	76.3 - *35.5*	*	127.7 126.8 *2.0*	*	130.6 129.6 *6.4*	-		*
3.(HF) (R,S)	*	48.9 - *22.2*	*	104.8 103.3 *8.2*	-		-		*
3.(FF) (R,S)	*	56.8 30.3 *23.9*	*	111.2 104.7 *0.07*	*	113.9 113.0 *1.15*	*	115.1 114.2 *5.5*	*
4.(HH) (R,S)	*	52.3; 56.7 - *4.5;* *20.2*	*	144.4 143.7 *1.7*	-		*	148.4 147.4 *6.8*	*
4.(HF) (R,S)	*	49.6; 60.2 - *25.9;* *3.2*	*	122.5 122.6 *6.2*	-		-		*

Table 2. *Cont.*

Acronym	Cr		SmC$_A$		SmC		SmA		Iso
5.(HH) (R,S)	*	60.4 - 35.6	*	136.5 135.9 1.1	-		*	143.1 142.2 6.9	*
5.(FH) (R,S)	*	69.8 - 33.6	*	108.1 95.6 0.04	*	119.6 118.8 1.1	*	129.2 127.8 6.3	*
5.(FF) (R,S)	*	72.6 14.3 25.2	*	110.9 101.6 0.03	*	119.8 118.9 0.9	*	126.9 125.7 5.6	*
6.(FH) (R,S)	*	59.8 - 25.9	*	120.8 120.3 0.75	-		*	131.5 130.5 6.5	*
6.(FF) (R,S)	*	68.4 40.4 26.9	*	121.3 120.4 0.76	-		*	129.1 127.6 5.9	*
7.(HH) (R,S)	*	36.0 - 13.25	*	114.3 105.5 0.03	*	131.2 130.3 0.8	*	140.1 139.5 6.8	*

First row—temperatures from DSC measurements obtained in the heating cycle; second row—temperatures from DSC measurements obtained in the cooling cycle; third row—transition enthalpies.

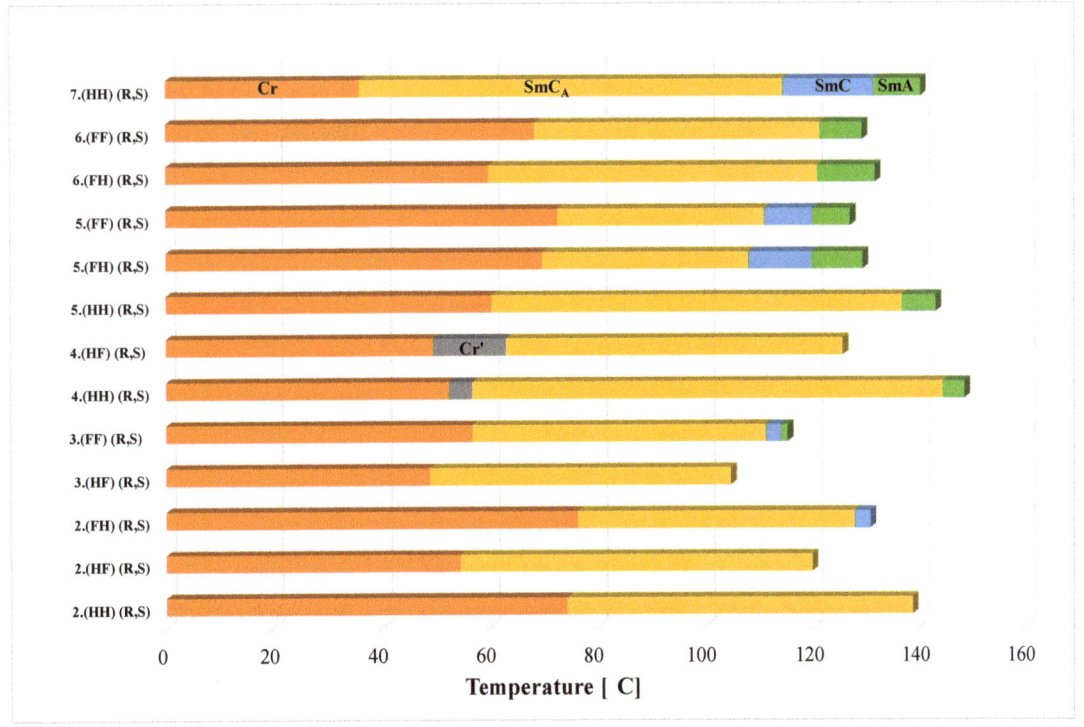

Figure 3. The temperature ranges of the phase transitions (from DSC measurements) for the racemic mixtures observed in the heating cycle.

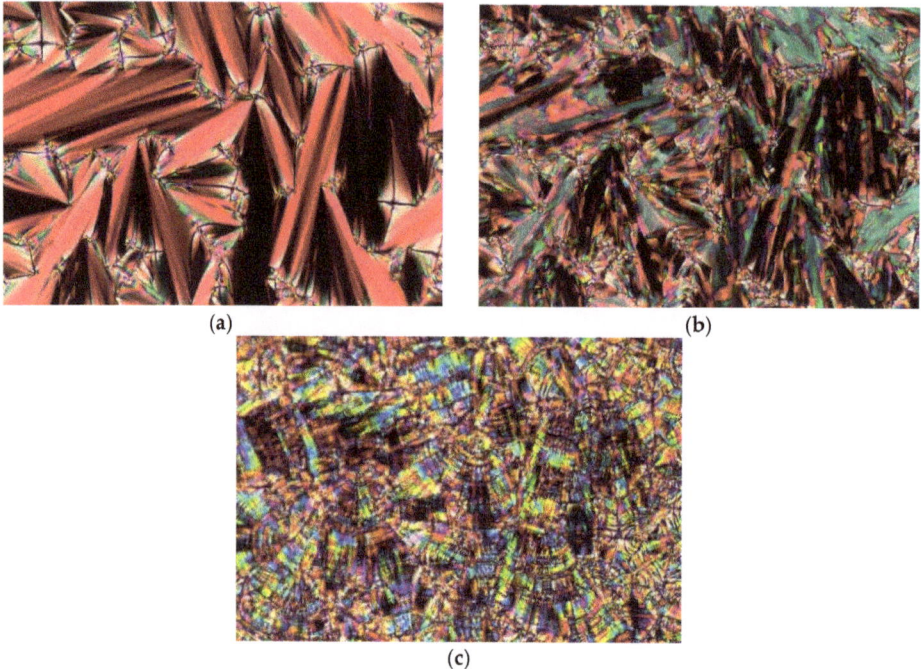

Figure 4. The textures of the smectic phases obtained in the cooling cycle for the racemic mixture **7.(HH) (R,S)**. (**a**) The SmA phase at 140.5 °C; (**b**) The SmC phase at 130.7 °C; (**c**) The SmC$_A$ phase at 98.2 °C.

All racemates have the anticlinic smectic phase in a broad temperature range. The broadest range of this phase is observed for the unsubstituted racemate with the oligomethylene spacer equal to 4 (above 87 °C). For four of the racemates, the direct transition from the SmC$_A$ phase to the isotropic phase is observed. Such properties occur for the racemates with a substitution of the (**HF**) and (**HH**) type and a short oligomethylene spacer ($n = 2, 3, 4$). It has been found that the direct SmC$_A$-Iso phase transition in the liquid crystalline materials is beneficial for improving the alignment in the electro-optical cell [33–35]. The SmC and SmA phases are observed in a short or medium temperature range. The unsubstituted racemates have the highest clearing points. The melting points change irregularly; the lowest temperature is observed for the racemate with the longest oligomethylene spacer (36 °C).

5. Mixtures Compositions and Their Properties

Six racemic mixtures were used as dopants to formulate new multicomponent antiferroelectric mixtures. For the six prepared doped mixtures, the temperatures and enthalpies of the phase transitions, as well as the helical pitch, were examined. The three base mixtures selected for the study, with the acronyms **W-458**, **W-459** and **W-460**, differ in the number of each doping components. All these mixtures have already been examined for their mesomorphic and thermodynamic properties, as well as the helical pitch [20,21]; therefore, it was possible to analyze the influence of the racemic mixtures on their properties. The base mixtures are characterized by very high values of the tilt angle of the molecules above 40° at lower temperatures. Six new mixtures with the racemic dopants at 20 wt% concentration were prepared (see Table 3).

Table 3. The compositions of the six prepared mixtures.

Acronyms of the Doped Mixtures	Base Mixtures	Acronyms of the Dopants
W-458A	W-458	2.(FH) (R,S)
W-458B	W-458	3.(HF) (R,S)
W-459A	W-459	5.(FF) (R,S)
W-459B	W-459	7.(HH) (R,S)
W-460A	W-460	4.(HF) (R,S)
W-460B	W-460	6.(FH) (R,S)

The phase transition temperatures for the prepared mixtures are summarized in Table 4 and visualized in Figure 5 (comparing them with the base mixtures).

Table 4. The mesomorphic and thermodynamic properties of the prepared mixtures. ("*" means that the phase is observed and "-" means that it is not observed; The enthalpy is indicated by the italics).

Mixtures	Cr	SmC$_A$*		SmC*		SmA*		Iso
W-458A	*	-	*	70.9–73.6 48.2–49.6 71.0 58.5 0.08	*	95.0–95.6 94.5–94.9 93.8 94.1 1.98	*	100.5–101.7 99.8–101.3 99.4 99.9 4.3	*
W-458B	*	-	*	74.8–76.0 57.2–57.8 75.0 63.8 0.15	*	92.8–93.2 92.5–92.8 91.8 92.0 2.15	*	97.0–97.7 96.4–97.1 96.2 96.3 4.4	*
W-459A	*	-	*	70.5–71.7 55.4–56.3 72.2 60.5 0.08	*	85.9–90.4 84.5–87.6 85.1 86.5 7.5	-		*
W-459B	*	-	*	72.5–73.6 61.7–62.5 73.1 64.9 0.75	*	86.4–92.4 84.8–89.8 84.9 88.2 6.1	-		*
W-460A	*	-	*	78.5–79.2 71.2–72.0 78.0 72.0 0.17	*	86.3–91.5 83.8–88.4 84.2 87.2 6.4	-		*
W-460B	*	-	*	72.0–73.1 63.1–61.9 71.3 63.8 0.07	*	84.0–84.7 83.2–84.0 82.3 83.4 1.63	*	88.7–93.2 87.3–90.6 87.2 89.6 4.1	*

First row—temperatures from POM measurements obtained in the heating cycle; Second row—temperatures from POM measurements obtained in the cooling cycle; Third row—temperatures from DSC measurements obtained in the heating cycle; Fourth row—temperatures from DSC measurements obtained in the cooling cycle; Fifth row—transition enthalpies [J·g^{-1}].

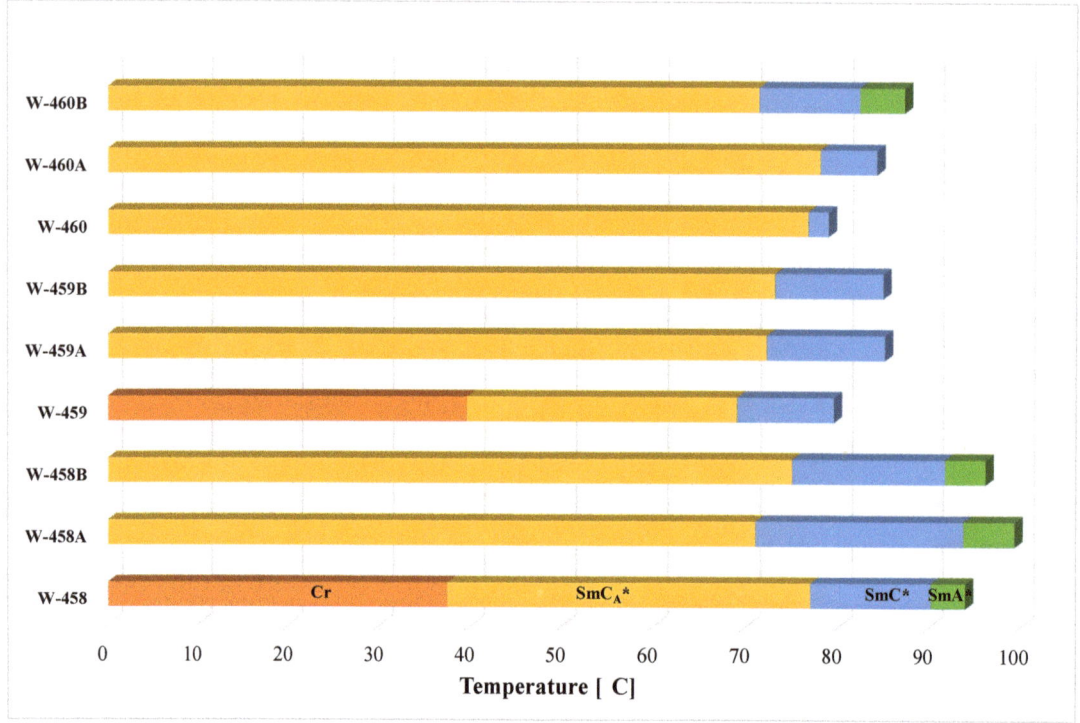

Figure 5. Comparing the temperature ranges of the phase transitions from DSC measurements for the base mixtures and the doped mixtures. ("*" means that the phase is observed).

All doped mixtures have the antiferroelectric phase (SmC$_A$*) in a very broad temperature range. The antiferroelectric phase range exceeds 70 °C for each of the doped mixtures. The broadest range of this phase is shown by the mixtures **W-458B** and **W-460A** doped with the racemic mixtures with the direct SmC$_A$-Iso phase transition with the acronyms **3.HF (R,S)** and **4.HF (R,S)**. The ferroelectric phase (SmC*) is present in all doped mixtures in a medium or a broad temperature range. The broadest range of this phase is shown by the mixtures **W-458A** and **W-458B** (23 and 17 degrees Celsius, respectively). The smectic A* phase occurs in a short temperature range in the mixtures **W-458A**, **W-458B** and **W-460B**.

All doped mixtures have higher clearing points than the base mixtures. None of the doped mixtures crystallized, and the DSC measurements were carried out from −15 °C.

The comparison of the helical pitch versus temperature for the base mixtures and the doped mixtures is shown in Figure 6a–c.

As could be expected (and of benefit), the doped mixtures are characterized by a longer helical pitch than the base mixture in both chiral phases. In the antiferroelectric phase, the helical pitch increases with increasing temperature, while in the ferroelectric phase, the helical pitch practically does not change upon heating, and it is very short (below 250 nm). The helical pitch in the SmC$_A$* phase reaches maximum values above 1400 nm for the mixtures **W-460A** and **W-460B**, while for the mixture **W-460B**, it occurs at a lower temperature.

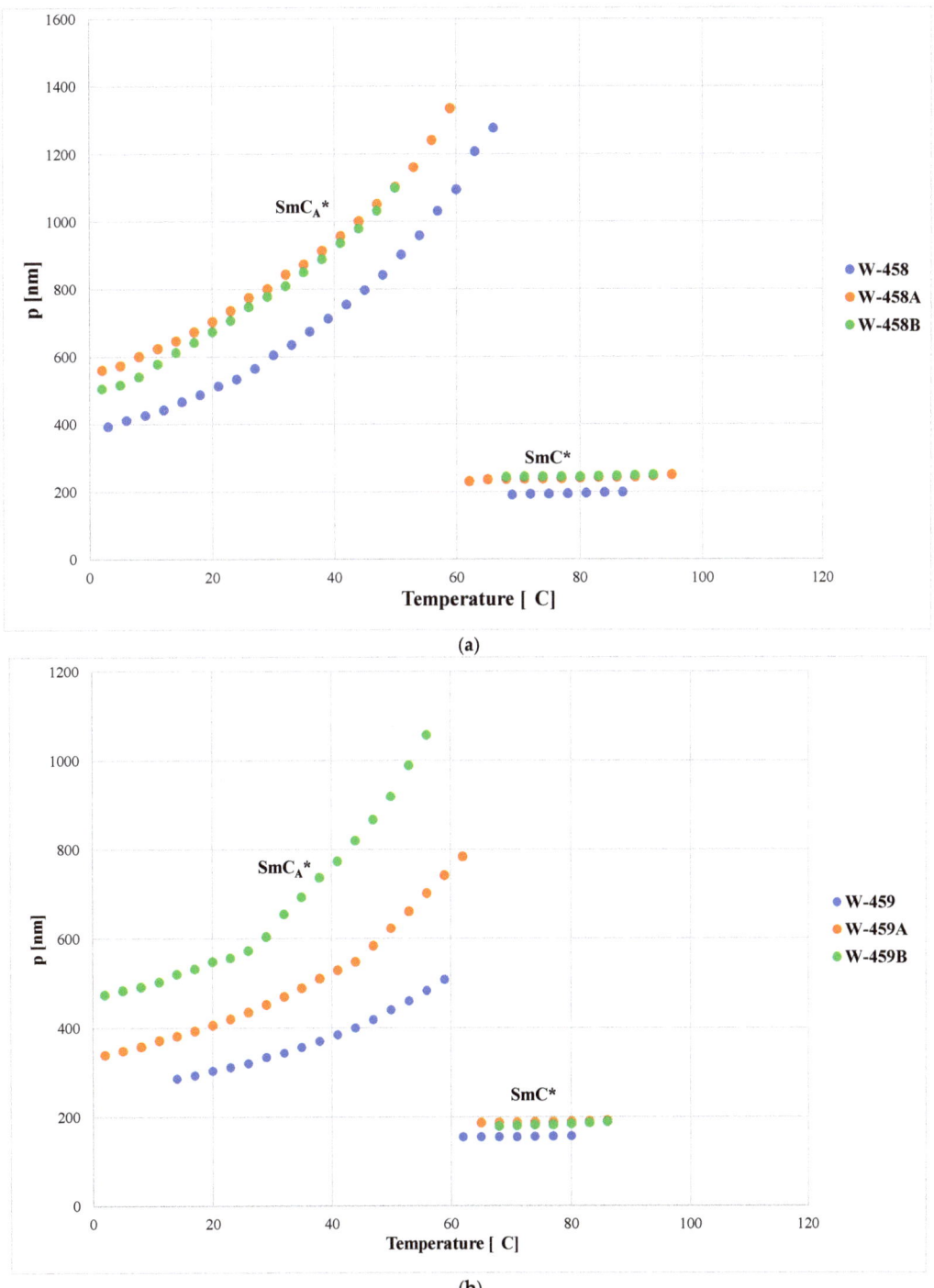

Figure 6. Cont.

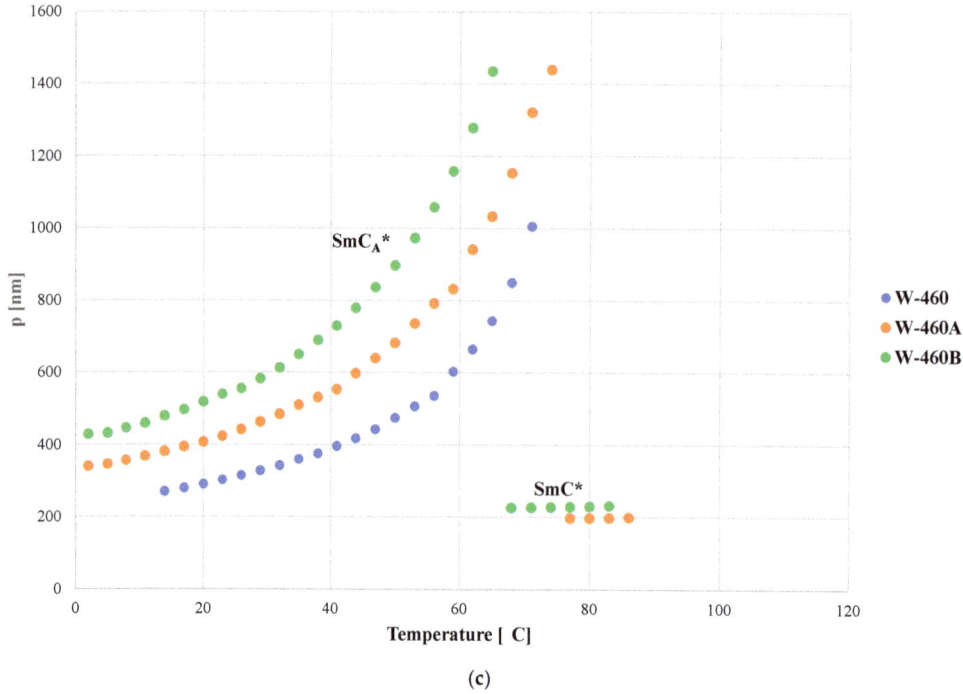

(c)

Figure 6. (a) The temperature dependence of the helical pitch for mixtures **W-458**, **W-458A** and **W-458B**. (b) The temperature dependence of the helical pitch for mixtures **W-459**, **W-459A** and **W-459B**. (c) The temperature dependence of the helical pitch for mixtures **W-460**, **W-460A** and **W-460B**. ("*" means that the phase is observed).

6. Conclusions

The newly synthesized racemates exhibit a wide temperature range for the anticlinic smectic C_A phase and are promising components of the liquid crystalline mixtures. The synclinic smectic C phase and smectic A phase in these racemates depend on the length of the oligomethylene spacer and the substitution of the benzene ring. The widest temperature range of the SmC_A phase is observed for the unsubstituted racemic mixtures.

The direct SmC_A-Iso transition is observed for the racemic mixtures with a short oligomethylene spacer: **2.(HH) (R,S)**, **2.(HF) (R,S)**, **3.(HF) (R,S)** and **4.(HF) (R,S)**. The SmC and SmA phases occur in a short or medium temperature range. New racemic mixtures can also be used as dopants which increase the helical pitch values and extend the temperature range of the SmC_A^* phase in the highly tilted antiferroelectric mixtures.

The mixtures doped with the racemates have favorable physical and electro-optical properties, as previously shown in Refs. [3–5]. In the next step, the measurements of the tilt angle of the molecules and the spontaneous polarization for the doped mixtures will be carried out.

It is also planned to develop the compositions of the other antiferroelectric liquid crystalline mixtures, based on the synthesized racemates, and separate the racemates into enantiomers by chiral liquid chromatography [36–40].

Supplementary Materials: The following supporting information can be downloaded at: https://www.mdpi.com/article//10.3390/cryst12121710/s1, Figures S1–S26: The NMR spectra of all racemic mixtures.

Author Contributions: Conceptualization, M.U.; methodology, M.U.; software, M.U.; resources, M.U. and M.S.; investigation, M.U. and M.S.; data curation, M.U.; writing-original draft preparation, M.U.; writing-review and editing, M.U. All authors have read and agreed to the published version of the manuscript.

Funding: This research was funded by the National Science Center Grant—the Miniatura 5 entitled "Optimization of the separation of liquid crystalline racemic mixtures on chiral columns by high performance liquid chromatography" (DEC-2021/05/X/ST4/0016) and University Research Grant (UGB 796/2022).

Data Availability Statement: The data presented in this study are available on request from the corresponding author.

Acknowledgments: The authors want to thank Bartłomiej Dębicki for his help in the measurements for the doped mixtures.

Conflicts of Interest: The authors declare no conflict of interest.

References

1. Gautier, R.; Klingsporn, J.M.; Van Duyne, R.P.; Poeppelmeier, K.R. Optical activity from racemates. *Nat. Mater.* **2016**, *15*, 591–592. [CrossRef] [PubMed]
2. Karnik, A.; Hasan, M. *Stereochemistry. A Three-Dimensional Insight*, 1st ed.; Elsevier: Amsterdam, The Netherlands, 2021.
3. Urbańska, M.; Dziaduszek, J.; Strzeżysz, O.; Szala, M. Synclinic and anticlinic properties of (R,S) 4'-(1-methylheptyloxycarbonyl) biphenyl-4-yl 4-[7-(2,2,3,3,4,4,4-heptafluorobutoxy)heptyl-1-oxy]benzoates. *Phase Transit.* **2019**, *92*, 657–666. [CrossRef]
4. Morawiak, P.; Żurowska, M.; Piecek, W. A long-pitch orthoconic antiferroelectric mixture modified by isomeric and racemic homostructural dopants. *Liq. Cryst.* **2018**, *45*, 1451–1459. [CrossRef]
5. Ogrodnik, K.; Perkowski, P.; Raszewski, Z.; Piecek, W.; Żurowska, M.; Dabrowski, R.; Jaroszewicz, L. Dielectric measurements of orthoconic antiferroelectric liquid crystal mixtures. *Mol. Cryst. Liq. Cryst.* **2011**, *547*, 54–64. [CrossRef]
6. Urbańska, M.; Morawiak, P.; Czerwiński, M. Effect of doping by enantiomers with the different absolute configuration and phase sequence on mesomorphic, helical and electro-optical properties of highly tilted chiral anticlinic mixture. *J. Mol. Liq.* **2020**, *309*, 113141. [CrossRef]
7. Czerwiński, M.; Tykarska, M.; Kula, P. New ferroelectric liquid crystalline materials with properties suitable for surface stabilized and deformed helix effects. *Liq. Cryst. Appl.* **2021**, *21*, 61–73. [CrossRef]
8. Nepal, S.; Das, B.; Das, M.K.; Das Sarkar, M.; Urbańska, M.; Czerwiński, M. Static permittivity and electro-optical properties of bi-component orthoconic antiferroelectric liquid crystalline mixtures targeted for polymer stabilized sensing systems. *Polymers* **2022**, *14*, 956. [CrossRef] [PubMed]
9. Chakraborty, S.; Das, M.K.; Bubnov, A.; Weissflog, W.; Węgłowska, D.; Dabrowski, R. Induced frustrated twist grain boundary liquid crystalline phases in binary mixtures of achiral hockey stick-shaped and chiral rod-like materials. *J. Mater. Chem. C* **2019**, *7*, 10530–10543. [CrossRef]
10. Tykarska, M.; Kurp, K.; Zieja, P.; Herman, J.; Stulov, S.; Bubnov, A. New quaterphenyls laterally substituted by methyl group and their influence on the self-assembling behaviour of ferroelectric bicomponent mixtures. *Liq. Cryst.* **2022**, *49*, 821–835. [CrossRef]
11. Debnath, A.; Mandal, P.K. Effect of fluorination on the phase sequence, dielectric and electro-optical properties of ferroelectric and antiferroelectric mixtures. *Liq. Cryst.* **2017**, *44*, 2192–2202. [CrossRef]
12. Vojtylová, T.; Kaspar, M.; Hamplová, V.; Novotná, V.; Sýkora, D. Chiral HPLC for a study of the optical purity of new liquid crystalline materials derived from lactic acid. *Phase Transit.* **2014**, *87*, 758–769. [CrossRef]
13. Vaňkátová, P.; Šrolerová, T.; Kubíčková, A.; Kalíková, K. Fast UHPLC enantioseparation of liquid crystalline materials with chiral center based on octanol in reversed-phase and polar organic mode. *Mon. Für Chem. Chem. Mon.* **2020**, *151*, 1235–1240. [CrossRef]
14. Vaňkátová, P.; Kalíková, K.; Kubíčková, A. Ultra-performance supercritical fluid chromatography: A powerful tool for the enantioseparation of thermotropic fluorinated liquid crystals. *Anal. Chim. Acta* **2018**, *1038*, 191–197. [CrossRef] [PubMed]
15. Urbańska, M.; Vaňkátová, P.; Kubíčková, A.; Kalíková, K. Synthesis, characterisation and supercritical fluid chromatography enantioseparation of new liquid crystalline materials. *Liq. Cryst.* **2020**, *47*, 1832–1843. [CrossRef]
16. Vojtylová-Jurkovičová, T.; Vaňkátová, P.; Urbańska, M.; Hamplová, V.; Sýkora, D.; Bubnov, A. Effective control of optical purity by chiral HPLC separation for ester-based liquid crystalline materials forming anticlinic smectic phases. *Liq. Cryst.* **2020**, *48*, 43–53. [CrossRef]
17. Vaňkátová, P.; Kubíčková, A.; Cigl, M.; Kalíková, K. Ultra-performance chromatographic methods for enantioseparation of liquid crystals based on lactic acid. *J. Supercrit. Fluids* **2019**, *146*, 217–225. [CrossRef]
18. Vaňkátová, P.; Kubíčková, A.; Kalíková, K. Enantioseparation of liquid crystals and their utilization as enantiodiscrimination materials. *J. Chromatogr. A* **2022**, *1673*, 463074. [CrossRef]
19. Vojtylová, T.; Hamplová, V.; Galewski, Z.; Korbecka, I.; Sýkora, D. Chiral separation of novel diazenes on a polysaccharide-based stationary phase in the reversed-phase mode. *J. Sep. Sci.* **2017**, *40*, 1465–1469. [CrossRef] [PubMed]

20. Urbańska, M.; Morawiak, P.; Senderek, M. Investigation of the tilt angle and spontaneous polarisation of antiferroelectric liquid crystals with a chiral centre based on (S)-(+)-3-octanol. *J. Mol. Liq.* **2021**, *328*, 115378. [CrossRef]
21. Urbańska, M.; Perkowski, P.; Morawiak, P.; Senderek, M. Antiferroelectric and ferroelectric mesophases created by (S) enantiomers with a short oligomethylene spacer and their usefulness in the formulation of orthoconic mixtures. *J. Mol. Liq.* **2020**, *320*, 114452. [CrossRef]
22. Sage, I. Thermochromic liquid crystals. *Liq. Cryst.* **2011**, *38*, 1551–1561. [CrossRef]
23. Chełstowska, A.; Czerwiński, M.; Tykarska, M.; Bennis, N. The influence of antiferroelectric compounds on helical pitch of orthoconic W-1000 mixture. *Liq. Cryst.* **2014**, *41*, 812–820. [CrossRef]
24. Czerwiński, M.; Tykarska, M.; Dabrowski, R.; Chełstowska, A.; Żurowska, M.U.; Kowerdziej, R.; Jaroszewicz, L.R. The influence of structure and concentration of cyano-terminated and terphenyl dopants on helical pitch and helical twist sense in orthoconic antiferroelectric mixtures. *Liq. Cryst.* **2012**, *39*, 1498–1502. [CrossRef]
25. Czerwinski, M.; Tykarska, M. Helix parameters in bi- and multicomponent mixtures composed of orthoconic antiferroelectric liquid crystals with three ring molecular core. *Liq. Cryst.* **2014**, *41*, 850–860. [CrossRef]
26. Żurowska, M.; Dąbrowski, R.; Dziaduszek, J.; Garbat, K.; Filipowicz, M.; Tykarska, M.; Rejmer, W.; Czupryński, K.; Spadło, A.; Bennis, N.; et al. Influence of alkoxy spacer length and fluorosubstitution of benzene ring on mesogenic and spectral properties of high tilted antiferroelectric 4'-(S)-1-(methylheptyloxycarbonyl)biphenyl-4-yl-4-(2,2,3,3,4,4,4-heptafluorobutoxy)alkoxybenzoates. *J. Mater. Chem.* **2010**, *21*, 2144–2153. [CrossRef]
27. Drzewiński, W.; Dąbrowski, R.; Czupryński, K. Orthoconic antiferroelectrics. Synthesis and mesomorphic properties of optically active (S)-(+)-4-(1-methylheptyloxycarbonyl)phenyl 4'-(fluoroalkanoyloxyalkoxy)biphenyl-4-carboxylates and 4'-(alkanoyloxyalkoxy)biphenyl-4-carboxylates. *Pol. J. Chem.* **2002**, *76*, 273–284. [CrossRef]
28. Belyakov, V.A.; Vladimir, E.D.; Orlov, V.P. Optics of cholesteric liquid crystals. *Sov. Phys. Uspekhi* **1979**, *22*, 64–88. [CrossRef]
29. Tykarska, M.; Czerwiński, M.; Miszkurka, J. Influence of temperature and terminal chain length on helical pitch in homologue series$n$H6Bi. *Liq. Cryst.* **2010**, *37*, 487–495. [CrossRef]
30. Tykarska, M.; Czerwiński, M. The inversion phenomenon of the helical twist sense in antiferroelectric liquid crystal phase from electronic and vibrational circular dichroism. *Liq. Cryst.* **2016**, *43*, 462–472. [CrossRef]
31. Raszewski, Z.; Kędzierski, J.; Perkowski, P.; Piecek, W.; Rutkowska, J.; Kłosowicz, S.; Zieliński, J. Refractive indices of the MHPB(H)PBC and MHPB(F)PBC antiferroelectric liquid crystals. *Ferroelectrics* **2002**, *276*, 289–300. [CrossRef]
32. Kowiorski, K.; Kędzierski, J.; Raszewski, Z.; Kojdecki, M.A.; Chojnowska, O.; Garbat, K.; Miszczyk, E.; Piecek, W. Complementary interference method for determining optical parameters of liquid crystals. *Phase Transit.* **2016**, *89*, 403–410. [CrossRef]
33. Milewska, K.; Drzewiński, W.; Czerwiński, M.; Dąbrowski, R.; Piecek, W. Highly tilted liquid crystalline materials possessing a direct phase transition from antiferroelectric to isotropic phase. *Mater. Chem. Phys.* **2016**, *171*, 33–38. [CrossRef]
34. Czerwiński, M.; Urbańska, M.; Bennis, N.; Rudquist, P. Influence of the type of phase sequence and polymer-stabilization on the physicochemical and electro-optical properties of novel high-tilt antiferroelectric liquid crystalline materials. *J. Mol. Liq.* **2019**, *288*, 111057. [CrossRef]
35. Czerwiński, M.; de Blas, M.G.; Bennis, N.; Herman, J.; Dmochowska, E.; Otón, J.M. Polymer stabilized highly tilted antiferroelectric liquid crystals—The influence of monomer structure and phase sequence of base mixtures. *J. Mol. Liq.* **2020**, *327*, 114869. [CrossRef]
36. Chankvetadze, B. Recent trends in preparation, investigation and application of polysaccharide-based chiral stationary phases for separation of enantiomers in high-performance liquid chromatography. *TrAC Trends Anal. Chem.* **2019**, *122*, 115709. [CrossRef]
37. Chen, X.; Yamamoto, C.; Okamoto, Y. Polysaccharide derivatives as useful chiral stationary phases in high-performance liquid chromatography. *Pure Appl. Chem.* **2007**, *79*, 1561–1573. [CrossRef]
38. Lämmerhofer, M. Chiral recognition by enantioselective liquid chromatography: Mechanisms and modern chiral stationary phases. *J. Chromatogr. A* **2010**, *1217*, 814–856. [CrossRef] [PubMed]
39. Scriba, G.K.E. Chiral recognition in separation science—An update. *J. Chromatogr. A* **2016**, *1467*, 56–78. [CrossRef]
40. Ward, T.J.; Ward, K.D. Chiral separations: A review of current topics and trends. *Anal. Chem.* **2011**, *84*, 626–635. [CrossRef]

Review

Development and Prospect of Viewing Angle Switchable Liquid Crystal Devices

Le Zhou *,† and Sijie Liu †

School of Materials Science and Engineering, Tsinghua University, Beijing 100084, China
* Correspondence: zhoule@pku.edu.cn; Tel.: +86-18811728321
† These authors contributed equally to this work.

Abstract: As we move from the industrial age to the information age, nowadays, the opportunity to access personal information in public increases as personal computers (PCs), cell phones, automated teller machines (ATM) and other portable display devices have come into wider use, so it is well suited for these liquid crystal displays (LCDs) to switch between wide viewing angle (WVA) (share mode) and narrow viewing angle (NVA) (privacy mode). In this review, we have summarized structures, principles and characteristics of several devices that show great potential application in controllable anti-peeping displays in the eyesight of materials consist of pure liquid crystals (LCs), polymer dispersed LCs (PDLCs), polymer stabilized LCs (PSLCs) or polymer network LCs (PNLCs) and non-LCs, which provides systematic information for next-generation viewing angle-controllable LCDs with lower operating voltage, higher transmittance and good viewing angle controllable characteristics. Because LCs/polymer composite films have the advantages of long life, low power consumption and energy saving, they are regarded as the mainstream technology of next-generation viewing angle controllable displays.

Keywords: wide viewing angle; narrow viewing angle; liquid crystals; PDLC; PSLC; PNLC

Citation: Zhou, L.; Liu, S. Development and Prospect of Viewing Angle Switchable Liquid Crystal Devices. *Crystals* **2022**, *12*, 1347. https://doi.org/10.3390/cryst12101347

Academic Editors: Yuriy Garbovskiy and Zhenghong He

Received: 28 August 2022
Accepted: 21 September 2022
Published: 24 September 2022

Publisher's Note: MDPI stays neutral with regard to jurisdictional claims in published maps and institutional affiliations.

Copyright: © 2022 by the authors. Licensee MDPI, Basel, Switzerland. This article is an open access article distributed under the terms and conditions of the Creative Commons Attribution (CC BY) license (https://creativecommons.org/licenses/by/4.0/).

1. Introduction

In this new digital information age, recently, with the rapid development of LCD technologies such as a wide view-twisted nematic (TN), in-plane switching (IPS), fringe-field switching (FFS), multidomain vertical alignment (MVA) and patterned vertical alignment (PVA), an increasing number of electronic products such as computer screens, mobile phones, electronic books, personal digital assistants (PDAs), tablet devices and LC televisions have been extensively applied [1–3]. Viewing angles are defined by a range of angles where the image displayed on an LCD remains suitable to the users, which refers to contour of isocontrast ratio that is dominated by the dark state dependent on the observation angle [1]. Ambient contrast ratio (ACR) is vital for evaluating an LCD performance, Figure 1a describes the schematic diagram of an LCD, the main reflection happens at front surface of LCD is defined as R_1, the ambient light passing into LCD is mostly absorbed by polarizers and optical components, by assuming no reflected light, $ACR_{LCD}(\theta, \phi) = \frac{L_{on}(\theta, \phi) + R_1}{L_{off}(\theta, \phi) + R_1}$, L_{on} and L_{off} represent the on-state and off-state luminance values of an LCD, θ and ϕ represent the polar angle and azimuthal angle, respectively [4]. In the past decades, WVA has been one of the most important characteristics in attaining the image quality in all viewing angle directions, which can be extended to 170° (polar angle) [5]. Currently, in applications of information devices in privacy contexts, such as using a notebook in public or an ATM machine, devices for adjusting viewing angle have drawn a lot of interest. Two different modes are demanded in the controllable viewing angle devices: share mode and privacy mode, generally, 3M light control films (also named anti-peeping films) including two films with a plurality of light absorbing regions are often used and taped on the front of screens (Figure 1b), the images on the LCD are visible only within a viewing angle range of −30° to 30° (Figure 1c), if a share mode is needed, the film must be removed from the LCD, which is unpractical and inconvenient [6]. Light control film containing closely spaced micro-louvers can also be LC or non-LC based, for LC-based materials, a lot of works focus on dual-mode switching of pure LC panel with alignment for viewing angle control [7]. To achieve both WVA and NVA properties in LCDs, various methods have been proposed for viewing angle control, such as multiple LC layers, one is utilized for gray level control, the other one is used for viewing angle control, which increases panel thickness and fabrication cost [1–7]. Dual backlight is introduced into viewing angle controllable devices, a normal backlight is for WVA mode and collimated backlight is for viewing angle control [1–7]. Dual cell method is proposed by using an additional LC cell to control the viewing angle, sub-pixel method in which the pixel is used to two parts, the main pixel is utilized for presenting images, sub-pixel is for controlling the viewing angle, which decreases aperture ratio of the main pixel [1–6]. Three-terminal electrode structure based on the FFS mode LCDs is introduced, different dark states are induced by various bias voltages, thus viewing angle controllable ability is presented [1–6]. However, these devices can achieve the viewing angle controllable characteristic, but have increase in thickness, cost and power consumption [1–7].

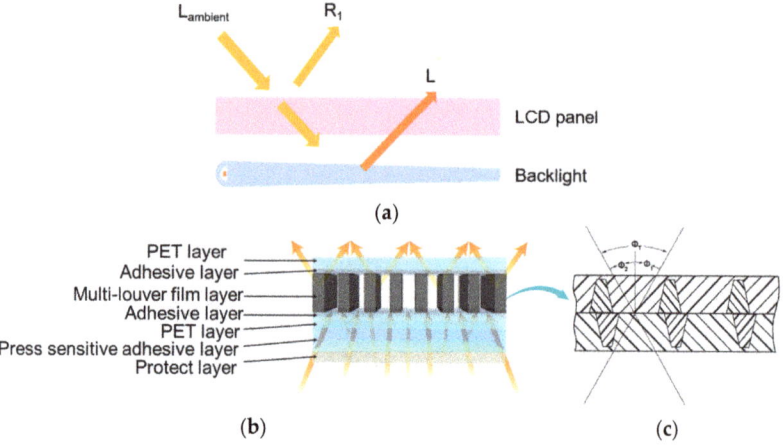

Figure 1. (a) Schematic diagram of an LCD; (b) Structure of 3M light control film; (c) The microlouver structure of 3M light control film.

Some researchers pay attention to light scattering ability of LC and polymer composite films, when the PDLC, PNLC or PSLC is at transparent state, the light remains narrow resulting in a NVA of the LCD. As the incident light transmits through PDLC, PNLC or PSLC, light is scattered because of the difference of refractive index between LCs and polymer, the LCD acts as a diffusing backlight source leading to a WVA. For non-LC devices, a multi-axial viewing control film for smart devices utilizing an array of optical micro-rods is proposed, it is convenient for users without removing the film from the information devices, additionally, a viewing angle switchable backlight module consists of a hybrid light guide plate and dual-light sources or two stacked backlight units is introduced [8,9]. However, switchable viewing angle displays based on non-LC devices require additional components or materials, thus cost, weight and thickness enhance.

In this review, we have summarized various designs to control the viewing angle and compared their characteristics in contrast ratios and viewing angles in WVA and NVA modes, taking thickness, brightness, power consumption, operating voltage and viewing angle switchable properties into consideration, LCs/polymer composite films show great potential for applications in viewing angle controllable LCDs.

2. Pure LC Devices

For pure LC system, methods for switchable viewing angle devices are divided into hybrid aligned nematic (HAN) LC cells, double cells, pixel division and three-electrode structure [10–15]. The controllable viewing angle device is located on the top of the LC panel, the device can be in transmissive mode (Figure 2a) or reflective mode (Figure 2b) [14]. To functionalize the device, the LC cell controls the transmission of light at predetermined azimuth angles and polar angles by applying the electric field on the LC layer, meanwhile, the light transmission in the normal direction is unchanged, thus the LC cell has no phase difference in the normal direction when changing the voltages [14]. For solving that, two types of LC cells including of homogenous aligned LC cell (Figure 2c) and HAN LC cell (Figure 2d) are utilized in this device [14]. The absorption axis of the polarizing films is parallel to the alignment of LC in both cells, in the homogeneous cell, the light-shielding effect exists in small range at an azimuth angle of $0°$, while a negative C-plate is added between the LC layer and the polarizing film in the HAN LC cell for obtaining the desired light polarization state shift, a wider range of light-shielding angles can be achieved [14].

In the MVA or PVA devices, in the dark state, to improve the brightness uniformity in the gray levels, the addition of a compensation film is for suppressing light leakage. In the film compensation method, to compensate for the dark state, a negative C plate and positive A plate are introduced for compensating for the dark state, a WVA with the contrast ratio over 10 is extended from $30°$ to $80°$ of polar angle is achieved [16]. The double-cell display is mainly composed of a vertically aligned LC layer for displaying information, a homogeneous aligned LC layer replacing the positive A plate for viewing angle switching, and a negative C plate for compensation under crossed polarizer [16]. Without electric field, the vertically aligned LCs with negative dielectric anisotropy tilt down to make an angle of $45°$, the optic axis of the homogeneous aligned LC layer with positive dielectric anisotropy is parallel to the analyzer's transmission axis, thus light transmitting through negative C plate and double LC cells is blocked by the analyzer, additionally, light leakage can be suppressed by homogeneous aligned LC cell and negative C plate [16]. For viewing angle switching, the mid-director of the homogeneous aligned LC layer is controlled by a vertical electric field (Figure 2e) [16], the high image quality in share mode has a polar viewing angle of $30°$ in the horizontal direction. However, this approach for tuning viewing angle requires additional components, leading to higher thickness and higher production cost.

When there is no electric field applied, blue phase typically appears in a very narrow temperature range, with polymer or nanomaterials stabilization, the temperature range can be extended, cubic structure in a BPLC appears to be optically isotropic, if there is a strong electric field, the anisotropy is induced along the electric field direction, which is defined as Kerr effect. By dividing pixel of the LCD device filled with a BPLC into two parts: a main pixel and a sub-pixel (Figure 2f–g), while the main pixel displays the image contents that are insensitive to viewing angle, as the birefringence is induced by transversal electric field and the LC reorientation is in the same plane, the sub-pixel controls the viewing angle. The LC cell is sandwiched between crossed polarizers, when applying a voltage between two electrodes, in-plane and vertical electric fields are produced in the main pixels and sub-pixels, respectively [16,17]. Without a bias voltage, BPLC is alike optically isotropic sphere, which is presented in Figure 2f, by the application of a bias voltage, the birefringence is induced due to Kerr effect, optic axes of BPLCs are parallel and perpendicular to the substrate in the main and sub-pixels, respectively (Figure 2g) [16,17]. A WVA is realized as the main pixels are only operated, the LCDs present a good dark in all directions, while a NVA is achieved as a bias voltage is applied to the sub-pixels, light leakage controlled by the applied voltage happens in the oblique viewing direction (Figure 2g) [16,17]. For WVA mode, the viewing angle at contrast ratio of over 10 exists up to $\sim 50°$ in all azimuthal directions, while for NVA mode, the viewing angle at contrast ratio of 10 exist along $20°$ in the horizontal and the vertical directions [10]. This approach exhibits some disadvantages such as lower contrast ratio or higher driving voltage and reduces transmittance more as the dark state is kept at the area of sub-pixel [18].

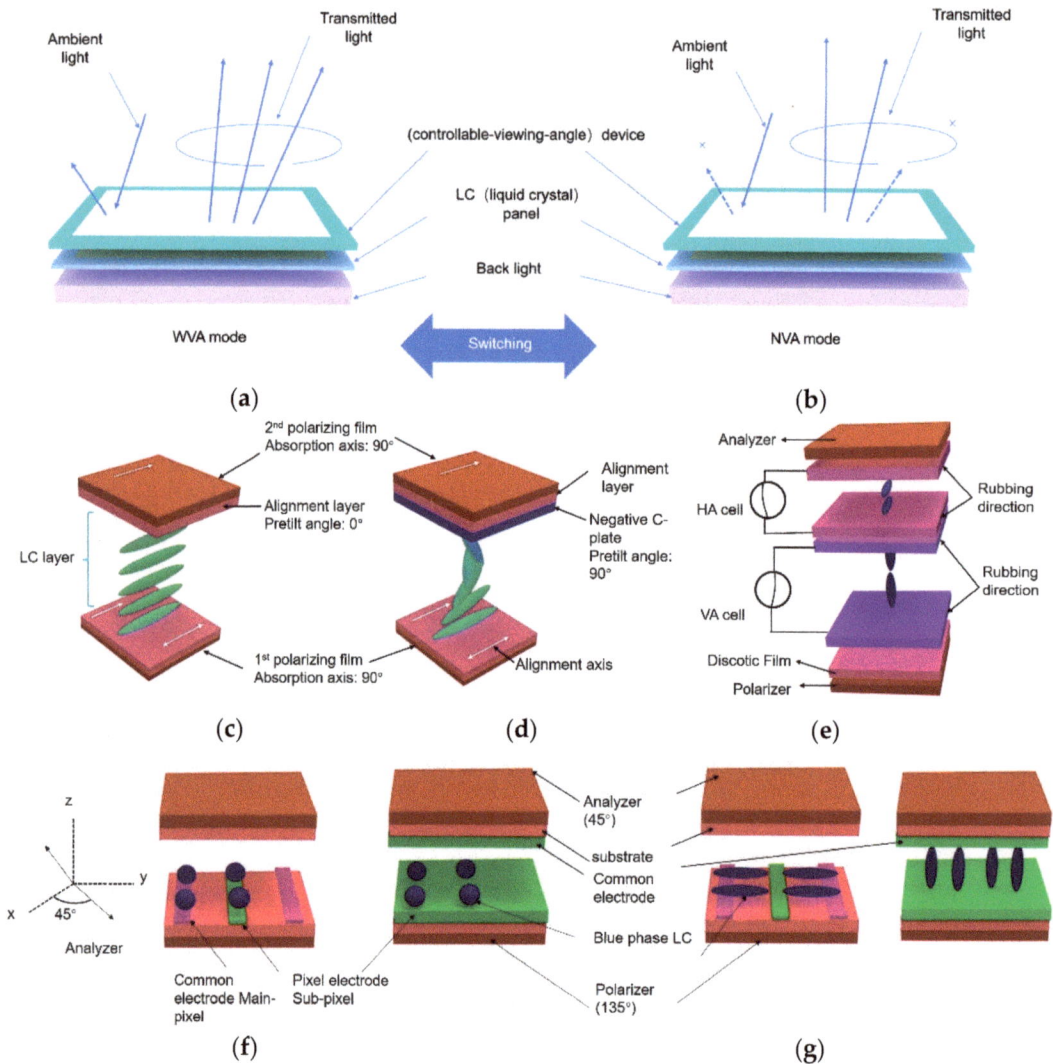

Figure 2. Schematic diagram of controllable viewing angle LCD: (**a**) WVA mode; (**b**) NVA mode. Schematic diagram of simulation models of controllable viewing angle device: (**c**) homogeneous aligned LC cell and (**d**) HAN LC cell; (**e**) optical cell configurations of the viewing angle device; (**f**) switchable LCD with an optically isotropic LC: voltage-off state and (**g**) voltage on state.

To decrease the volume, weight and fabrication cost of LCD, a viewing angle switchable LCD owning an interlayer support with surface microstructures that is placed between the top and bottom substrates, the original LC layer is for presenting images and the complementary LC layer is for viewing angle tunning, thus WVA and NVA can be switched between ± 70° and ± 40° [19]. A viewing angle controllable LCD using two stable states of bistable nematic LC such as splay and 180°-twist at π cell with three-terminal electrodes, WVA is shown at splay state with interdigitated electrodes at the IPS or FFS modes, NVA is presented at the twisted state with vertical electrode [20].

For a controllable viewing angle LCD inserting by BPLC cell, BPLC plays as a tunable positive C-film or a negative C-film, the LCD has no alignment layer and sub-millisecond response time [21]. By designing electrode structure that is same as the one used for dual-mode switching in FFS LCD, an additional electrode is introduced to the top substrate, viewing angle with a contrast ratio of over 10 is tuned between $\pm 70°$ and $\pm 10°$ along the azimuthal direction by applying a small vertical bias voltage [22]. By using in-plane electrodes and etched substrate, double in-plane electric field is for reducing the driving voltage, operating voltage decreases to 8.2 V [18,23]. A fringe and in-plane switching BPLCD with a top common electrode has been proposed [24], of which three-terminal electrode structure gives it good dark state for NVA display and similar voltage-dependent transmission curves for NVA and WVA displays. If there is no bias voltage on the top common electrode, it shows the similar viewing angle, contrast ratio and transmission compared to that of conventional BPLCD. If different various bias voltages are applied on the top common electrode, viewing angle can continuously and uniformly change from WVA to NVA at a high contrast ratio of over 500 [25,26]. However, electrode structure designing method leads to a gray inversion in NVA mode utilizing vertical and horizontal fields at the same time.

By adopting a thermally variable retardation layer (TVRL) composed of a homeotropically aligned nematic LC with transparent electric-heating lines to control temperature, the LCD shows continuous and symmetric viewing angle characteristics, by thermally changing the birefringence of the film, the WVA mode is obtained by Joule heating in the TVRL where the LCs are isotropic phase, while the NVA mode is achieved at the nematic phase in the TVRL, the polar angle with the contrast ratio of 10 decreases up to 20° along the diagonal axis [27–29]. Except for these approaches, by using a guest-host (GH) LC cell doped with 5.0 wt.% of a dichroic dye (DD), as the polarization direction is orthogonal to the long molecular axis of DD, light transmits the cell, if the polarization is parallel to the long molecular axis of DD, light is absorbed by DD in the cell, the NVA and WVA modes can be realized by applying the voltage of 10.0 V across the GH cell [30].

3. PDLC Devices

Utilizing 3M light-control film and PDLC film together to create switchable anti-peeping film, PDLC is a voltage-controlled LC film that can switch between transparent and scattering (Figure 3a), light scattering intensity of PDLC film is dependent on electric field, switchable voltage between light scattering (share mode) and transparent state (privacy mode) is approximately 5.0 V [31]. As 3M light control film and PDLC film are not integrated, a control method for light distribution patterns of PDLCs is established by controlling an internal polymer structure formed by irradiating unidirectional diffused UV light in isotropic phase (Figure 3b) [32]. To investigate the effect of micro-louver size on the viewing angle controllable ability, four micro-louver structures with the same thickness are fabricated by utilizing thiol-ene photopolymerization (40-40 μm, 40-60 μm, 60-40 μm and 60-60 μm) (Figure 3c–f), four controllable anti-peeping devices can realize WVA and NVA modes at 0 and 8 V [33]. However, external switchable anti-peeping films need to be installed on the screen, to solve this problem, the LCD containing the switchable anti-peeping film is fabricated based on PDLC, the propagation direction of the light passing the 3M light control film can be tuned by the PDLC film, the NVA mode is presented without electric field and WVA mode is shown in the application of electric field (24 V) [34].

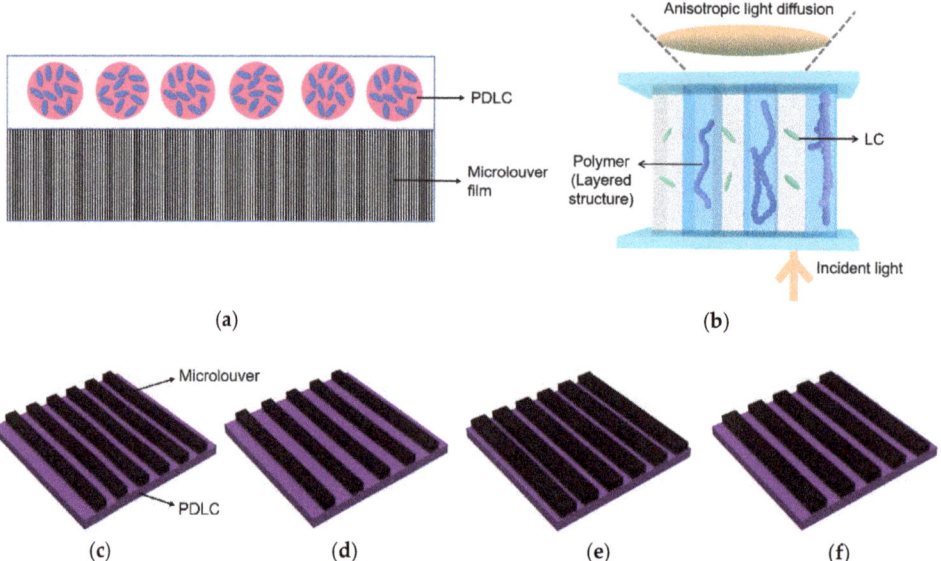

Figure 3. (**a**) Configuration structure and schematic mechanism of switchable anti-peeping film consists of 3M light control film and PDLC film; (**b**) schematic representation of PDLC having layered distribution of polymer network structure; (**c–f**) four microlouver structures with the same thickness, (**c**), 40–40 μm, (**d**), 40–60 μm, (**e**), 60–40 μm, (**f**), 60–60 μm, the first number represents the width of micro-stripe and the second number represents the space between two microstripes.

4. PSLC/PNLC Devices

The structure of the viewing angle-tunable LCD is depicted in Figure 4a, it consists of two cells: an ultra-super twisted (UST) cell, which is situated between the upper polarizer and the TN-TFT cell, and a PNLC cell, which is situated between the lower polarizer and the backlight. When the electric field is absent, the PNLC cell acts as a negative retarder, causing viewing angle to be wider than a conventional TN-cell [35]. Polymer stabilized BP (PSBP) LCD becomes more and more attractive due to no alignment, the simple fabrication process and fast response time, a PSBP LCD with double-size IPS electrode structure is proposed, which shows transflective characteristics because the bottom electrodes are made by aluminum material. With the application of a bias voltage on the top electrodes, a good viewing angle controllable display is exhibited [36]. A viewing angle-controllable LCD utilizing PSBPLC and a single-panel method, without electric field, a BPLC appears to be optically isotropic, if there is a strong electric field, due to Kerr effect, the birefringence is induced along the electric field direction in the BPLC and increases with the electric field, viewing angle of the device can be controlled from 100° to 30° (Figure 4b) [37]. We have developed an electrically switchable viewing angle device that fabricated by DDs doped polymer stabilized cholesteric LCs (DD-PSCLCs) [38], with the additional photo-mask in the UV irradiation process, polymer has formed in the irradiated part, as shown in Figure 4c, when applying a relative lower electric field, CLCs in the irradiated part turn to be in focal conic state, while CLCs in the non-LC-irradiated part still keep planar state, as is described in Figure 4d, if the electric field is higher, CLCs in the irradiated part change to be homeotropic state, while CLCs in the irradiated part still be in the planar state (Figure 4e).

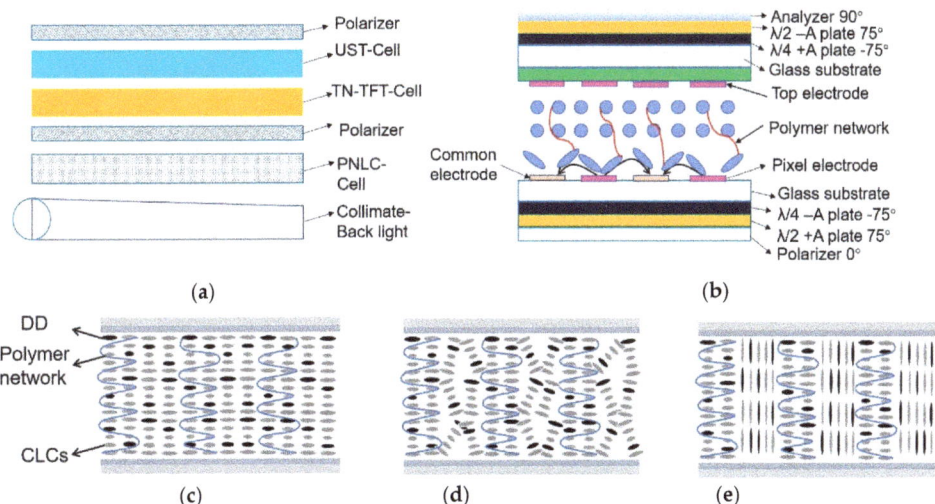

Figure 4. (**a**) Structure of a viewing angle controllable device; (**b**) Cell structure of the proposed viewing angle-controllable PSBP-LCD; (**c**–**e**) Various states of CLCs in the DD-PSCLC device under the application of electric field: (**c**) planar state; (**b**) focal conic state; (**e**) homeotropic state.

5. Non-LC Devices

Setting an array of optical micro-rods, consisting of electrophoretic material that composed of black particles dispersed in a transparent medium, with a rectangular parallelepiped shape, whose size is 40 μm wide, 40 μm deep and 120 μm high, while the space between each optical micro-rod is 10 μm [8,39]. As the black particles are gathered electronically to one side in the transparent resin, the cross-stripe part filled by transparent substrate transmit incident light the same as the light-transmitting portions. A viewing angle switching device based on array of optical micro-rod is demonstrated (Figure 5a), as the black particles which enables switching two states between a WVA and a NVA in one second at 20 V of applied voltage (Figure 5b). Two modes are electrically switched, (a) NVA mode and (b) WVA mode. The limited viewing angle mode is +/− 30° of the visible angle and 50% of the transmittance, and the one for the non-limited viewing mode is 58% of the transmittance by the applied voltage of 20 V. A compact light guide plate (LGP) with special designed microstructures and dual light sources is proposed (Figure 5c), of which the micro-prisms are utilized to guide the emitting rays toward normal viewing cones while a set of strip-patterned diffuser toward wide viewing cones [40,41]. By simply switching the separate light sources, quick switching between the two modes is possible. If light source 1 is on, the backlight operates in the NVA mode, and if light source 2 is on, the backlight operates in the WVA mode [9]. Figure 5d represents the viewing angle switchable backlight module that is divided into three parts, the hybrid light guide plate (HLGP) is called the light incidence surface, the first light sources provide NVA mode and the second light sources provide WVA mode, Figure 5e depicts side view of the viewing angle controllable backlight module, the HLGP is composed of main layer and sub-layer with various refractive indices, the top micro-prisms are spaced with a graded interval for controlling the uniformity degree and the bottom micro-prisms reflect the rays to the main layer [9]. Consequently, in the NVA mode, the half-luminance angle decreases 11 degrees in the horizontal direction, while in the WVA mode, the half-luminance angle increases 56 degrees in the horizontal direction [9].

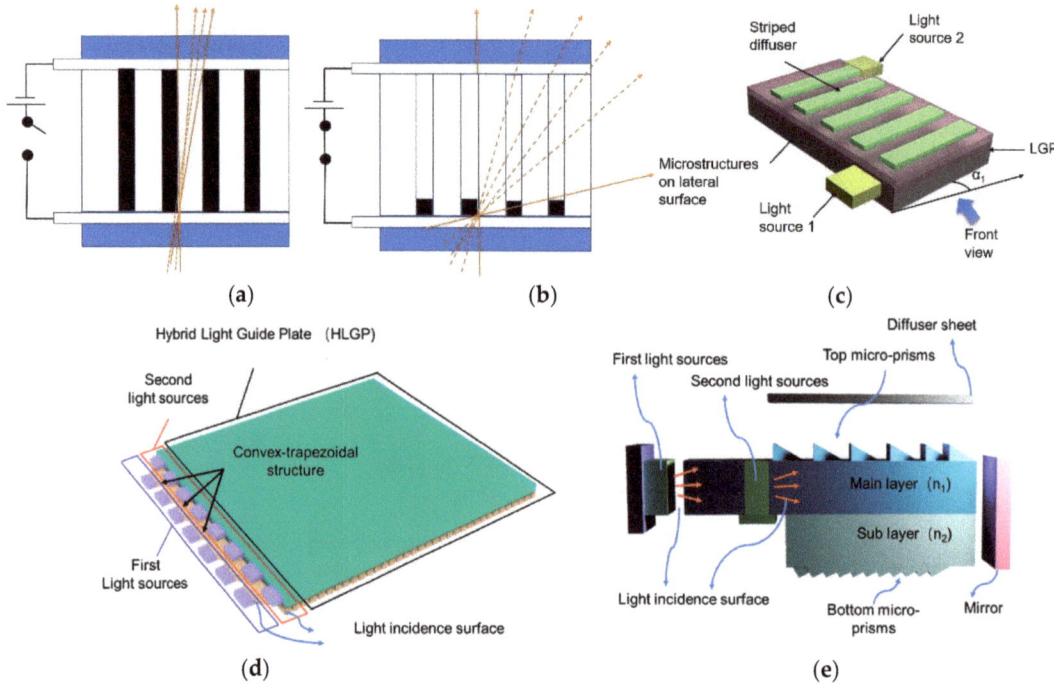

Figure 5. Illustration of concept for (**a**) narrow mode and (**b**) wide mode can be selected electronically. The narrow mode (**a**) is achieved as black particles are fully dispersed, and the wide mode (**b**) is achieved as black particles are completely gathered one side in the optically transparent medium. (**c**) Sketch of the backlight configuration, (**d**,**e**) optical structure of the dual backlight viewing angle switchable device: (**d**) tilt view; (**e**) side view.

6. Discussion

Over the past few decades, viewing angle controllable devices for LCDs have been introduced, comparative analysis of performances of various LC and non-LC devices is presented in Table 1, dual cell method increases the production cost for needing two LC cells, two-pixel method reduces transmittance more as the dark state is held at the area of sub-pixel, but viewing angle in the narrow mode is only 20°, three-terminal electrode structure based on the FFS mode LCDs has simple structure, but the contrast ratio for NVA is not good because of light leakage induced by the bias voltage. The existing switchable privacy protection displays based on LC and non-LC devices (Table 2) on the market have been applied in ATMs, monitors and high-end notebooks, but the function of switchable anti-peeping function leads to higher power consumption or lower image quality, moreover, the switchable viewing angle effect is mainly in horizontal view.

To consider next-generation commercial switchable anti-peeping products, LC devices especially for LC/polymer composite films may be applied to almost any situations due to their tunable optical properties and easier fabrication, for example, by achieving various controllable microlouver structures under the mask-assisted UV curing of LC monomers, DD-doped PSCLC device can be developed to meet thinness, low power consumption, high image quality, it is believed that this new viewing angle switching mode will have strong potential for future display applications.

Table 1. Comparison of the performances of various LC and non-LC devices.

Types	Principles	WVA/ Contrast Ratio	WVA/ Viewing Angle	NVA/ Contrast Ratio	NVA/ Viewing Angle	References
LC devices	Pixel division	>10:1	50°	10:1	20°	10
	Optically isotropic LC	/	120° polar angle	/	35°	11
	Homogeneous aligned LC layer	10:1	170°	2:1	60°	16
	Electrical tilting of LC	10:1	±70°	10:1	±10°	21
	TVRL	10:1	80° polar angle	10:1	20°	28
	3M/PDLC	/	±60°	/	±30°	30
	Microlouver/PDLC	/	±62°	/	±39°	32
	PSBPLC	/	100°	/	30°	37
Non-LC devices	Dual light source	/	140°	/	60°	9
	Optical micro-rods	/	/	/	±30°	38
	Striped diffuser	/	±55°	/	±10°	41

Table 2. Comparison of various LC and non-LC devices.

Types	Privacy Effect	Privacy Viewing Angle	Thickness	Power	Application
LC devices	Excellent	Horizontal perspective	Thin	High	Monitors
Non-LC devices	Moderate	Horizontal perspective	Thick	Moderate	ATMs

7. Conclusions and Perspective

In conclusion, it is impractical to preserve privacy in display devices by adding more panels and backlight units due to the increasing panel thickness and high production costs. Utilizing a dual lighting system is also not recommended for achieving NVA due to its greater cost and insufficient brightness, a more expensive LCD is required for developing electrode structures. The electronically switchable scattering and transparent modes of polymer and LCs composite films allow the incident light to be either weakly (strongly) scattered or transmitted. Due to their simple manufacturing process, it is possible to produce perfect viewing angle controllable commercial LCDs by inserting the polymer/LCs composite films into the backlight, adding some special patterns or gratings to the films during the photopolymerization process of UV cured LC or non-LC monomers under masks, controlling their thickness and microstructures, contrast ratio between scattering mode and transparent mode and driving voltage can be both further tuned, moreover, multi-directional switchable viewing angle devices are also developed by utilizing the system of polymer/LCs composites.

Recently, viewing angle controllable devices have been the mainstream of high-end displays, while the width of microlouvers in the devices is still difficultly to be refined. Additionally, the brightness of backlight is relatively low in the WVA mode compared to that in the NVA mode. Moreover, the current anti-peeping effect originated from the NVA mode in the devices is mainly demonstrated in horizontal view, to meet the needs of all-angle anti-peeping effect in the future displays, the horizontal and vertical viewing angle controllable devices are the development trend. Adjustable viewing angle devices are currently popular in shielding sensitive or private information, except for that, they can also avoid the light of PCs from interfering with others or reduce the light on the car display screens that disturbs the drivers at night.

Author Contributions: L.Z. and S.L. contributed equally to this work; writing—review and editing. All authors have read and agreed to the published version of the manuscript.

Funding: This research received no external funding.

Institutional Review Board Statement: Not applicable.

Informed Consent Statement: Not applicable.

Data Availability Statement: Not applicable.

Conflicts of Interest: The authors declare no conflict of interest.

References

1. Chen, C.P.; Jhun, C.G.; Yoon, T.H.; Kim, J.C. Viewing angle switching of tristate liquid crystal display. *Jpn. J. Appl. Phys.* **2007**, *46*, L676–L678. [CrossRef]
2. Chen, C.P.; Kim, K.H.; Yoon, T.H.; Kim, J.C. A viewing angle switching panel using guest-host liquid crystal. *Jpn. J. Appl. Phys.* **2009**, *48*, 062401. [CrossRef]
3. Her, J.H.; Shin, S.J.; Lim, Y.J.; Bhattacharyya, S.S.; Kang, W.S.; Lee, G.D.; Lee, S.H. Viewing angle switching in in-plane switching liquid crystal display. *Mol. Cryst. Liq. Cryst.* **2011**, *544*, 220–226. [CrossRef]
4. Chen, H.W.; Tan, G.J.; Wu, S.T. Ambient contrast ratio of LCDs and OLED displays. *Opt. Express* **2017**, *25*, 33643–33656. [CrossRef]
5. Jo, S.I.; Lee, S.G.; Lee, Y.J.; Kim, J.H.; Yu, C.J. Viewing angle controllable liquid crystal display under optical compensation. *Opt. Eng.* **2011**, *50*, 094003. [CrossRef]
6. Chiu, R.C. Light Control Device. U.S. Patent 6398370, 4 June 2002.
7. Baek, J.I.; Kim, K.H.; Kim, J.C.; Yoon, T.H. Viewing angle control of a hybrid-aligned liquid crystal display. *Mol. Cryst. Liq. Cryst.* **2009**, *498*, 103–109. [CrossRef]
8. Shiota, K.; Okamoto, M.; Tanabe, H. 76-3: Distinguished Paper: Viewing-angle-switching device based on array of optical micro-rod incorporated with electrophoretic material systems. *SID Symp. Dig. Tech. Pap.* **2017**, *48*, 1117–1120. [CrossRef]
9. Chen, B.T.; Pan, J.W.; Hu, Y.W.; Tu, S.H.P. 50: Design of a novel hybrid light guide plate for viewing angle switchable backlight module. *SID Symp. Dig. Tech. Pap.* **2013**, *44*, 1181–1184. [CrossRef]
10. Kim, M.S.; Lim, Y.J.; Yoon, S.; Kang, S.W.; Lee, S.H.; Kim, M.; Wu, S.T. A controllable viewing angle LCD with an optically isotropic liquid crystal. *J. Phys. D Appl. Phys.* **2010**, *43*, 145502. [CrossRef]
11. Kim, M.S.; Lim, Y.J.; Yoon, S.; Kim, M.K.; Kumar, P.; Kang, S.W.; Kang, W.S.; Lee, G.D.; Lee, S.H. Luminance-controlled viewing angle-switchable liquid crystal display using optically isotropic liquid crystal layer. *Liq. Cryst.* **2011**, *38*, 371–376. [CrossRef]
12. Lim, Y.J.; Jeong, E.; Kim, Y.S.; Jeong, Y.H.; Jang, W.G.; Lee, S.H. Viewing angle switching in fringe-field switching liquid crystal display. *Mol. Crystal. Liq. Cryst.* **2008**, *495*, 186. [CrossRef]
13. Lim, Y.J.; Kim, J.H.; Her, J.H.; Bhattacharyya, S.S.; Park, K.H.; Lee, J.H.; Kim, B.K.; Lee, S.H. Viewing angle controllable liquid crystal display with high transmittance. *Opt. Express* **2010**, *18*, 6824–6830. [CrossRef]
14. Adachi, M. Controllable-viewing-angle display using a hybrid aligned nematic liquid-crystal cell. *Jpn. J. Appl. Phys.* **2008**, *47*, 7920–7925. [CrossRef]
15. Adachi, M.; Shimura, M. P-228L: Late-News Poster: Controllable Viewing-Angle Displays using a Hybrid Aligned Nematic Liquid Crystal Cell. *SID Symp. Dig. Tech. Pap.* **2006**, *37*, 705–708. [CrossRef]
16. Jeong, E.; Lim, Y.J.; Rhee, J.M.; Lee, S.H.; Lee, G.D.; Park, K.H.; Choi, H.C. Viewing angle switching of vertical alignment liquid crystals by controlling birefringence of homogeneously aligned liquid crystal layer. *Appl. Phys. Lett.* **2007**, *90*, 051116. [CrossRef]
17. Jeong, E.; Lim, Y.J.; Chin, M.H.; Kim, J.H.; Lee, S.H.; Ji, S.H.; Lee, G.D.; Park, K.H.; Choi, H.C.; Ahn, B.C. Viewing-angle controllable liquid crystal display using a fringe- and vertical-field driven hybrid aligned nematic liquid crystal. *Appl. Phys. Lett.* **2008**, *92*, 261102. [CrossRef]
18. Li, Y.F.; Sun, Y.B.; Zhao, Y.L.; Li, P.; Ma, H.M. A continuous viewing angle controllable blue phase liquid crystal display. *J. Disp. Technol.* **2014**, *10*, 799–803. [CrossRef]
19. Kim, Y.T.; Hong, J.H.; Cho, S.M.; Lee, S.D. Viewing angle switchable liquid crystal display with double layers separated by an interlayer support. *Jpn. J. Appl. Phys.* **2009**, *48*, 110205. [CrossRef]
20. Gwag, J.S.; Lee, Y.J.; Kim, M.E.; Kim, J.H.; Kim, J.C.; Yoon, T.H. Viewing angle control mode using nematic bistability. *Opt. Express* **2008**, *16*, 2663–2669. [CrossRef]
21. Rao, L.H.; Ge, Z.B.; Wu, S.T. Viewing angle controllable displays with a blue-phase liquid crystal cell. *Opt. Express* **2010**, *18*, 3143–3148. [CrossRef]
22. Baek, J.I.; Kim, K.H.; Lee, S.R.; Kim, J.C.; Yoon, T.H. Viewing angle control of a fringe-field switching cell by electrical tilting of liquid crystal. *Jpn. J. Appl. Phys.* **2008**, *47*, 1615–1617. [CrossRef]

23. Sun, Y.B.; Li, Y.F.; Zhao, Y.L.; Li, P.; Ma, H.M. A low voltage and continuous viewing angle controllable blue phase liquid crystal display. *J. Disp. Technol.* **2014**, *10*, 484–487. [CrossRef]
24. Yu, Y.N.; Dou, H.; Ma, H.M.; Sun, Y.B. Continuous viewing angle controllable patterned vertical alignment liquid crystal display. *Liq. Cryst.* **2014**, *41*, 1595–1599. [CrossRef]
25. Yu, Y.N.; Dou, H.; Ma, H.M.; Sun, Y.B. Viewing angle controllable fringe and inplane switching vertical alignment LCD. *Liq. Cryst.* **2015**, *42*, 316–321. [CrossRef]
26. Hu, D.; Chen, M.; Li, D.; Yu, G.; Sun, Y.B. A controllable viewing angle optical film using micro prisms filled with liquid crystal. *Liq. Cryst.* **2021**, *48*, 1373–1381.
27. Gwag, J.S.; Han, I.Y.; Yu, C.J.; Choi, H.C.; Kim, J.H. Continuous viewing angle-tunable liquid crystal display using temperature-dependent birefringence layer. *Opt. Express* **2009**, *17*, 5426–5432. [CrossRef]
28. Gwag, J.S.; Lee, Y.J.; Han, I.Y.; Yu, C.J.; Kim, J.H. LCD with tunable viewing angle by thermal modulation of optical layer. *J. Inf. Disp.* **2009**, *10*, 19–23. [CrossRef]
29. Han, I.Y.; Gwag, J.S.; Lee, Y.J.; Yu, C.J.; Kim, J.H. Viewing angle controllable liquid crystal display by thermally variable retardation layer. *Mol. Cryst. Liq. Cryst.* **2009**, *507*, 122–128. [CrossRef]
30. Choi, H.J.; Lee, H.S.; Lim, S.H.; Park, S.Y.; Baek, S.K.; Lee, J.H. Dependence of the viewing angle control property of a guest-host liquid crystal cell on the extinction coefficient of the mixture. *Appl. Opt.* **2019**, *58*, 6105–6111. [CrossRef]
31. Zhou, L.; He, Z.M.; Han, C.; Zhang, L.Y.; Yang, H. Switchable anti-peeping film for liquid crystal displays from polymer dispersed liquid crystals. *Liq. Cryst.* **2019**, *46*, 718–724. [CrossRef]
32. Ishinabe, T.; Horii, Y.; Shibata, Y.; Fujikake, H. Structured PDLCs for controlling LCD viewing-angle. *Dig. Tech. Pap.* **2018**, *49*, 546–549. [CrossRef]
33. Han, C.; Zhou, L.; Ma, H.P.; Li, C.Y.; Zhang, S.F.; Cao, H.; Zhang, L.Y.; Yang, H. Fabrication of a controllable anti-peeping device with a laminated structure of microlouver and polymer dispersed liquid crystals film. *Liq. Cryst.* **2019**, *46*, 2235–2244. [CrossRef]
34. He, Z.M.; Shen, W.B.; Yu, P.; Zhao, Y.Z.; Zeng, Z.; Liang, Z.; Chen, Z.; Zhang, H.M.; Zhang, H.Q.; Miao, Z.C.; et al. Viewing-angle-switching film based on polymer dispersed liquid crystals for smart anti-peeping liquid crystal display. *Liq. Cryst.* **2022**, *49*, 59–65. [CrossRef]
35. Hisatake, Y.; Kawata, Y.; Murayama, A. Viewing angle controllable LCD using variable optical compensator and variable diffuser. *SID Sym. Dig. Tech. Pap.* **2005**, *36*, 1218–1221. [CrossRef]
36. Li, P.; Sun, Y.B.; Wang, Q.H. A transflective and viewing angle controllable blue-phase liquid crystal display. *Liq. Cryst.* **2013**, *40*, 1024–1027. [CrossRef]
37. Liu, L.W.; Cui, J.P.; Li, D.H.; Wang, Q.H. A viewing-angle-controllable blue-phase liquid-crystal display. *J. SID* **2012**, *20*, 337–340.
38. Yang, H.; Zhou, L.; Ma, H.; Han, C.; Hu, W.; Zhang, L.Y. Electric Control Dimming Film for Use in Display, Has First Composite Material Area Formed by Dye Molecules and Polymer Network, Where First and Second Composite Material Areas Are Formed in Material Film Layer along Vertical Direction. Chinese Patent CN201710456245.9, 6 September 2019.
39. Shiota, K.; Okamoto, M.; Tanabe, H. Viewing-angle-switching device based on array of optical micro-rods incorporated with electrophoretic material systems. *J. Soc. Inf. Disp.* **2017**, *25*, 76–82. [CrossRef]
40. Wang, Y.J.; Lu, J.G.; Chao, W.C. P-76: Distinguished Student Poster: Viewing Angle Switchable Display with a Compact and Directional Backlight Module. *SID Symp. Dig. Tech. Pap.* **2014**, *45*, 1270–1273. [CrossRef]
41. Wang, Y.J.; Lu, J.G.; Chao, W.C.; Shieh, H.P.D. Switchable viewing angle display with a compact directional backlight and striped diffuser. *Opt. Express* **2015**, *23*, 21443–21454. [CrossRef]

MDPI
St. Alban-Anlage 66
4052 Basel
Switzerland
www.mdpi.com

Crystals Editorial Office
E-mail: crystals@mdpi.com
www.mdpi.com/journal/crystals

Disclaimer/Publisher's Note: The statements, opinions and data contained in all publications are solely those of the individual author(s) and contributor(s) and not of MDPI and/or the editor(s). MDPI and/or the editor(s) disclaim responsibility for any injury to people or property resulting from any ideas, methods, instructions or products referred to in the content.